Uncertainty and Context in GIScience and Geography

T0225288

Uncertainty and context pose fundamental challenges in GIScience and geographic research. Geospatial data are imbued with errors (e.g., measurement and sampling) and various types of uncertainty that often obfuscate any understanding of the effects of contextual or environmental influences on human behaviors and experiences. These errors or uncertainties include those attributable to geospatial data measurement, model specifications, delineations of geographic context in space and time, and the use of different spatiotemporal scales and zonal schemes when analyzing the effects of environmental influences on human behaviors or experiences. In addition, emerging sources of geospatial big data – including smartphone data, data collected by GPS, and various types of wearable sensors (e.g., accelerometers and air pollutant monitors), volunteered geographic information, and/or location-based social media data (i.e., crowd-sourced geographic information) – inevitably contain errors, and their quality cannot be fully controlled during their collection or production.

Uncertainty and Context in GIScience and Geography: Challenges in the Era of Geospatial Big Data illustrates how cutting-edge research explores recent advances in this area, and will serve as a useful point of departure for GIScientists to conceive new approaches and solutions for addressing these challenges in future research. The seven core chapters in this book highlight many challenges and opportunities in confronting various issues of uncertainty and context in GIScience and geography, tackling different topics and approaches.

The chapters in this book were originally published as a special issue of the *International Journal of Geographical Information Science*.

Yongwan Chun is Associate Professor of Geospatial Information Sciences (GIS) at the University of Texas at Dallas, USA. His research interests lie in GIS and spatial statistical approaches to solving geographical problems, including geographic flow modeling, space-time modeling, and uncertainty.

Mei-Po Kwan is Choh-Ming Li Professor of Geography and Resource Management and Director of the Institute of Space and Earth Information Science at the Chinese University of Hong Kong, China. Her research interests include environmental health, human mobility, sustainable cities, urban, transport and social issues in cities, and GIScience.

Daniel A. Griffith is Ashbel Smith Professor of Geospatial Information Sciences at the University of Texas at Dallas, USA, and has authored numerous books and academic articles, garnering him many awards. He pursues research at the interface between geography and mathematics, especially statistics. His current research emphasizes visualization, space-time analysis, and public health.

Uncertainty and Context in GIScience and Geography

Challenges in the Era of Geospatial Big Data

Edited by
**Yongwan Chun, Mei-Po Kwan, and
Daniel A. Griffith**

Routledge
Taylor & Francis Group

LONDON AND NEW YORK

First published 2021
by Routledge
2 Park Square, Milton Park, Abingdon, Oxon OX14 4RN

and by Routledge
52 Vanderbilt Avenue, New York, NY 10017

Routledge is an imprint of the Taylor & Francis Group, an informa business

British Library Cataloguing in Publication Data
A catalogue record for this book is available from the British Library

ISBN13: 978-0-367-64299-0 (hbk)
ISBN13: 978-0-367-64300-3 (pbk)
ISBN13: 978-1-003-12384-2 (ebk)

Typeset in MyriadPro
by Newgen Publishing UK

Publisher's Note
The publisher accepts responsibility for any inconsistencies that may have arisen during the conversion of this book from journal articles to book chapters, namely the inclusion of journal terminology.

Disclaimer
Every effort has been made to contact copyright holders for their permission to reprint material in this book. The publishers would be grateful to hear from any copyright holder who is not here acknowledged and will undertake to rectify any errors or omissions in future editions of this book.

Contents

Citation Information

The following chapters were originally published in the *International Journal of Geographical Information Science*, volume 33, issue 6 (2019). When citing this material, please use the original page numbering for each article, as follows:

Chapter 1

Uncertainty in the effects of the modifiable areal unit problem under different levels of spatial autocorrelation: a simulation study
Sang-Il Lee, Monghyeon Lee, Yongwan Chun and Daniel A. Griffith
International Journal of Geographical Information Science, volume 33, issue 6 (2019), pp. 1135–1154

Chapter 2

Spatial autocorrelation and data uncertainty in the American Community Survey: a critique
Paul H. Jung, Jean-Claude Thill and Michele Issel
International Journal of Geographical Information Science, volume 33, issue 6 (2019), pp. 1155–1175

Chapter 3

Uncertainties in the geographic context of health behaviors: a study of substance users' exposure to psychosocial stress using GPS data
Mei-Po Kwan, Jue Wang, Matthew Tyburski, David H. Epstein, William J. Kowalczyk and Kenzie L. Preston
International Journal of Geographical Information Science, volume 33, issue 6 (2019), pp. 1176–1195

Chapter 4

Exploring the uncertainty of activity zone detection using digital footprints with multi-scaled DBSCAN
Xinyi Liu, Qunying Huang and Song Gao
International Journal of Geographical Information Science, volume 33, issue 6 (2019), pp. 1196–1223

Chapter 5

Same space – different perspectives: comparative analysis of geographic context through sketch maps and spatial video geonarratives

Andrew Curtis, Jacqueline W. Curtis, Jayakrishnan Ajayakumar, Eric Jefferis and Susanne Mitchell

International Journal of Geographical Information Science, volume 33, issue 6 (2019), pp. 1224–1250

Chapter 6

Travel impedance agreement among online road network data providers

Eric M. Delmelle, Derek M. Marsh, C. Dony and Paul L. Delamater

International Journal of Geographical Information Science, volume 33, issue 6 (2019), pp. 1251–1269

Chapter 7

A network approach to the production of geographic context using exponential random graph models

Steven M. Radil

International Journal of Geographical Information Science, volume 33, issue 6 (2019), pp. 1270–1288

For any permission-related enquiries please visit:
www.tandfonline.com/page/help/permissions

Notes on Contributors

Jayakrishnan Ajayakumar, GIS Health and Hazards Lab, Department of Geography, Kent State University, Ohio, USA.

Yongwan Chun, School of Economic, Political and Policy Sciences, University of Texas at Dallas, Richardson, TX, USA.

Andrew Curtis, Department of Population and Quantitative Health Sciences, School of Medicine, Case Western Reserve University, Ohio, USA.

Jacqueline W. Curtis, Department of Population and Quantitative Health Sciences, School of Medicine, Case Western Reserve University, Ohio, USA.

Paul L. Delamater, Department of Geography and Carolina Population Center, University of North Carolina at Chapel Hill, NC, USA.

Eric M. Delmelle, Department of Geography and Earth Sciences and Center for Applied GIS, University of North Carolina at Charlotte, NC, USA.

Coline Dony, American Association of Geographers, Washington, DC, NW, USA.

David H. Epstein, Treatment Section, Clinical Pharmacology and Therapeutics Research Branch, Intramural Research Program, National Institute on Drug Abuse, Baltimore, MD, USA.

Song Gao, Department of Geography, University of Wisconsin–Madison, WI, USA.

Daniel A. Griffith, School of Economic, Political and Policy Sciences, University of Texas at Dallas, Richardson, TX, USA.

Qunying Huang, Department of Geography, University of Wisconsin–Madison, WI, USA.

Michele Issel, Department of Public Health Sciences, University of North Carolina at Charlotte, NC, USA.

Eric Jefferis, College of Public Health, Kent State University, Ohio, USA.

Paul H. Jung, Department of Geography and Earth Sciences, University of North Carolina at Charlotte, NC, USA.

William J. Kowalczyk, Department of Psychology, Hartwick College, Oneonta, NY, USA.

Mei-Po Kwan, Department of Geography and Resource Management and Institute of Space and Earth Information Science, The Chinese University of Hong Kong, Shatin, Hong Kong.

Monghyeon Lee, Samsung Electronics Co. Ltd., Hwasung-si, Gyeonggi-do, South Korea.

Sang-Il Lee, Department of Geography Education, Seoul National University, Seoul, South Korea.

Xinyi Liu, Department of Geography, University of Wisconsin–Madison, WI, USA.

Derek M. Marsh, Department of Geography and Earth Sciences and Center for Applied GIS, University of North Carolina at Charlotte, NC, USA.

Susanne Mitchell, College of Public Health, Kent State University, Ohio, USA.

Kenzie L. Preston, Treatment Section, Clinical Pharmacology and Therapeutics Research Branch, Intramural Research Program, National Institute on Drug Abuse, Baltimore, MD, USA.

Steven M. Radil, Department of Geography, University of Idaho, Moscow, ID, USA.

Jean-Claude Thill, Department of Geography and Earth Sciences, University of North Carolina at Charlotte, NC, USA.

Matthew Tyburski, Treatment Section, Clinical Pharmacology and Therapeutics Research Branch, Intramural Research Program, National Institute on Drug Abuse, Baltimore, MD, USA.

Jue Wang, Department of Geography, University of Toronto–Mississauga, ON, Canada.

Introduction[1]

Yongwan Chun, Mei-Po Kwan, and Daniel A. Griffith

Uncertainty and context pose fundamental challenges in GIScience and geographic research. Geospatial data are imbued with errors (e.g., measurement, sampling, and stochastic noise) and various types of uncertainty that often obfuscate any understanding of the effects of contextual or environmental influences on human behaviors and experiences. These errors or uncertainties include those attributable to geospatial data measurement, model specifications, delineations of geographic context in space and time, and the use of different spatiotemporal scales and zonal schemes when analyzing the effects of environmental influences on human behaviors or experiences (Kwan 2012, 2018a, 2018b).

The literature still encompasses many gaps in uncertainty research in GIScience and geography, although considerable efforts have been devoted to conceptual and/or methodological developments to appropriately address these gaps. For example, Griffith et al. (2015) discuss pertinent concerns that need to be addressed in spatial analysis, including impacts of uncertainty on spatial patterns and models, uncertainties arising from spatial data aggregation (related to areal unit definitions), visualization of uncertainty, and metadata for data quality. In addition, emerging sources of geospatial big data – including smartphone data, data collected by global positioning systems (GPS) and various types of wearable sensors (e.g., accelerometers and air pollutant monitors), volunteered geographic information (VGI), and/or location-based social media data (i.e., crowdsourced geographic information) – inevitably contain errors, and their quality cannot be fully controlled during their collection or production. They almost always are noisy, messy, and dirty data (Sapountzi and Psannis 2020).

Further, the uncertain geographic context problem (Kwan 2012) recognizes challenges in identifying the "true causally relevant" spatial and temporal contexts that influence people's behaviors and experiences. That is, conventionally summarized values based on fixed and static areal units, especially those amorphous ones that are based on people's residential location or neighborhood, cannot be used to adequately identify the true geographic context or its true effects on individuals who navigate inhomogeneous and non-isotropic spaces over time. As recent studies have shown, ignoring people's daily mobility and exposures to nonresidential contexts may lead to erroneous results when assessing people's exposures to, and the health impacts of, environmental factors (Kwan 2018b).

---

1 Note that this introduction is a slightly modified version of the guest editors' editorial (Chun et al., 2019) appearing in a special issue of *International Journal of Geographical Information Science* (vol. 33, no. 6, pp. 1131–1288).

Advances in GIScience, especially in methods for collecting and analyzing big spatial data collected via personal sensing and location-based services, enable more accurate delineation of individual-level real-time contexts as well as assessment of people's exposures to these contexts. To generate reliable geographic knowledge, these uncertainties and contextual issues need to be addressed as part of GIScience and geographic research endeavors.

To address these uncertainties and contextual problems, and to enrich geographic knowledge, the 2017 American Association of Geographers (AAG) annual meeting organized and featured *Uncertainty and Context in Geography and GIScience* as one of its main special themes. The *International Journal of Geographical Information Science* (Chun et al. 2019) published a special issue as an extended effort to emphasize the importance of this theme, and to further foster advances in this research area. This book republishes the seven papers appearing in this special issue that cover three main topics. The first topic is uncertainty issues arising from well-known geographic concepts, addressing the modifiable areal unit problem (MAUP) and spatial autocorrelation (SA) complications. The second topic is impacts of the uncertainty in emerging new sources of geographic big data on geographic analysis; specifically, VGI for road network and geo-tagged tweet data are discussed here. The third topic is the uncertain geographic context problem (UGCoP), which is further investigated and addressed. As an introductory overview, the following section provides a brief summary of the seven papers republished in this book.

In Chapter 1, Lee et al. revisit and extend discussion about uncertainty issues related to the MAUP, focusing on how SA is convoluted with it. Although papers in the literature have shown that these fundamental phenomena in GIScience have an impact on each other, how they affect each other remains elusive. Unlike many contributions to the literature that utilize a set of empirical variables, their paper examines a wide range of SA values with Moran spatial weights matrix eigenvectors in an extensive experimental simulation. Their analysis results confirm that an initial SA level at the finest spatial scale makes a substantial difference to MAUP effects. They also explore how an index quantifying global SA, namely the Moran coefficient, behaves in terms of scale and zoning effects, although the existing convolution still is shown to make generalizing uncertainty patterns problematic.

Jung et al. argue in Chapter 2 that a popular SA measure, again the Moran coefficient, should be cautiously utilized when uncertainty is present in georeferenced observations. Whereas American Community Survey (ACS) data are widely adopted to explore neighborhood effects on a geographical phenomenon, uncertainty in ACS data that mostly arises from sampling error often is not recognized or incorporated into spatial analysis. The authors adjust the Moran coefficient to incorporate margins-of-error for ACS data based upon the framework of errors-in-variables independent variable measurement error conceptualizations, deriving its expected value and variance in doing so. They show that SA patterns can be substantially different when the uncertainty in ACS data is incorporated in an examination of teen birth rates in Mecklenburg County, North Carolina. This paper suggests that spatial patterns generated with ACS data can be misleading if their uncertainty is not accounted for.

Through a study of substance users' exposure to psychosocial stress, in Chapter 3, Kwan et al. examine how differently delineated contextual areas may lead to altered exposure estimates. This study uses GPS data collected from 47 outpatients with substance use disorders in Baltimore, Maryland, to assess their exposures to environmental stress based upon two variables: community socioeconomic status, and crime. It compares seven different methods for defining individual activity space using 35.2 million GPS tracking points collected from the participants. The results indicate that the different methods yield different exposure estimates, some of which may lead to different conclusions in studies using only one of the other methods. This investigation has important implications for future research about the effect of contextual influences on health behaviors and outcomes: whether or not a study observes any significant influence of an environmental factor on health may depend upon what contextual units are used to assess individual exposure.

In Chapter 4, Lin et al. investigate uncertainty in detecting activity zones of individuals using digital footprint data from a social media platform. Sources of errors in activity zone detection include the reliability of social media data, methodological limitations, and representations (or delineations) of activity zones. Focusing on a methodological limitation, these authors propose the multi-scaled extension of the popular density-based spatial clustering of applications with noise (DBSCAN) method, which commonly is criticized because of its sensitivity to parameter settings, but nevertheless is popularly utilized. They argue that this proposed method, which allows localized parameter settings for each activity zone, outperforms other popular detection methods, and can reduce noise considerably in detection results.

Through a case study of criminal ex-offenders, in Chapter 5, Curtis et al. examine how more meaningful and nuanced individual geographic contexts can be identified and incorporated into a spatial analysis based on qualitative and mixed methods. This study furnishes a comparison of the sketch maps and spatial video geonarratives (SVGs) of 11 ex-offenders that identify high crime areas of their communities. The results indicate that SVGs consistently help generate spatial data at finer scales, and identify more relevant locations than the sketch maps. SVGs also provide explanations of the spatial-temporal processes and causal mechanisms associated with specific places. The authors conclude that the use of SVGs can be a rigorous method for collecting data about the geographic context for many phenomena.

To assess the quality of VGI, Delmelle et al., in Chapter 6, evaluate the agreement in travel impedance between estimates from MapQuest Open that is based on OpenStreetMap (OSM) data, and estimates from two other popular commercial providers, namely Google Maps and ArcGIS Online. Using an artificial dataset for the road network of the state of North Carolina, this study simulates potential routes, estimates their travel impedance using a routing service Application Program Interface (API), and extracts the average number of OSM contributors for each route. The results reveal a strong correlation in travel impedance among the three road networks, and imply that travel impedance agreement is the greatest in areas with denser road networks, and the least for routes of shorter distances. This study concludes that larger groups of VGI contributors hold higher potential for validating and correcting inherent errors in VGI datasets.

To go beyond context as an abstraction, and to address the need to formally operationalize the notion of context for research purposes, in Chapter 7, Radil develops a multiscalar framework for use with social network-based statistical models called exponential random graph models (ERGMs). This framework emphasizes the importance of agency for both geographic and social contexts, while also recognizing place-specific and larger-scale influences. Using network data about World War I, the author generates a series of

ERGMs to demonstrate the importance of multiple types of contexts to a set of observed outcomes. This paper confirms the value of continued engagement with a wider range of theories and approaches pertaining to how and why context matters.

Although these seven papers address different topics using different approaches, they highlight many challenges and opportunities in addressing various issues of uncertainty and context in GIScience and geography. We hope this book illustrates how cutting-edge research explores recent advances in this area, and will serve as a useful point of departure for GIScientists to conceive new approaches and solutions for addressing these challenges in future research. Certainly, one emerging theme signaled by these papers is the need for a geography of uncertainty. Although global measures of uncertainty now accompany more contemporary maps, little attention has been paid to their corresponding geographic distributions. Geostatistics furnishes a possible prototype for moving forward: mean response maps supplemented by (perhaps in juxtaposition with) prediction error maps. The advent of ACS data enables this type of convention for many more geographic situations and phenomena. For example, Koo et al. (2018) provide a spatial data analysis tool capable of efficiently and effectively portraying uncertainty across a map, illustrating their tool with ACS uncertainty data. Another emerging theme portended by these papers is the importance of SA to geographic context. Although SA is recognized widely, especially by quantitative spatial scientists, as a fundamental property of georeferenced data, positive SA has monopolized their attention. Negative SA is the most neglected concept in spatial analysis (Griffith 2019). Furthermore, increasing evidence is accumulating that a mixture of positive and negative SA characterizes most geographic contexts (see Hu et al. 2020). In conclusion, the uncertainty and context topic should spawn a research agenda guiding numerous geospatial investigations for years to come. This book contributes to the foundation of that agenda.

References

Chun, Y., Kwan, M-P., and Griffith, D. (2019). Uncertainty and context in GIScience and geography: challenges in the era of geospatial big data. *International Journal of Geographical Information Science*, 33(6), 1131–1134.

Curtis, A., Curtis, J., Ajayakumar, J., Jefferis, E., and Mitchell, S. (2019). Same space – different perspectives: comparative analysis of geographic context through sketch maps and spatial video geonarratives. *International Journal of Geographical Information Science*, 33(6), 1224–1250.

Delmelle, E., Marsh, D., Dony, C., and Delamater, P. (2019). Travel impedance agreement among online road network data providers. *International Journal of Geographical Information Science*, 33(6), 1251–1269.

Griffith, D. (2019). Negative spatial autocorrelation: one of the most neglected concepts in spatial statistics. *Stats*, 2, 388–415.

Griffith, D., Wong, D., and Chun. Y. (2015). Uncertainty-related research issues in spatial analysis, in J. Shi, B. Wu, & A. Stein (Eds.), *Uncertainty modelling and quality control for spatial data*. London: Taylor & Francis Group/CRC Press, 3–11.

Hu, L., Chun, Y., and Griffith, D. (2020). Uncovering a positive and negative spatial autocorrelation mixture pattern: a spatial analysis of breast cancer incidences in Broward County, Florida, 2000–2010. *Journal of Geographical Systems*, 22, 291–308.

Jung, P., Thill, J.-C., and Issel, L. (2019). Spatial autocorrelation and data uncertainty in the American Community Survey: a critique. *International Journal of Geographical Information Science*, 33(6), 1155–1175.

Koo, H., Chun, Y., and Griffith, D. (2018). Integrating spatial data analysis functionalities in a GIS environment: spatial analysis using ArcGIS Engine and R (SAAR). *Transactions in GIS*, 22(3), 721–736.

Kwan, M.-P. (2012). The uncertain geographic context problem. *Annals of the Association of American Geographers*, 102(5), 958–968.

Kwan, M.-P. (2018a). The limits of the neighborhood effect: contextual uncertainties in geographic, environmental health, and social science research. *Annals of the American Association of Geographers*, 108(6), 1482–1490.

Kwan, M.-P. (2018b). The neighborhood effect averaging problem (NEAP): an elusive confounder of the neighborhood effect. *International Journal of Environmental Research and Public Health*, 15, 1841.

Kwan, M.-P., Wang, J., Tyburski, M., Epstein, D., Kowalczyk, W., and Prestonet, K. (2019). Uncertainties in the geographic context of health behaviors: a study of substance users' exposure to psychosocial stress using GPS data. *International Journal of Geographical Information Science*, 33(6), 1176–1195.

Lee, S.-I., Lee, M., Chun, Y., and Griffith, D. (2019). Uncertainty in the effects of the modifiable areal unit problem under different levels of spatial autocorrelation: a simulation study. *International Journal of Geographical Information Science*, 33(6), 1135–1154.

Liu, X., Huang, Q., and Gao, S. (2019). Exploring the uncertainty of activity zone detection using digital footprints with M-DBSCAN. *International Journal of Geographical Information Science*, 33(6), 1196–1223.

Radil, S. (2019). A multiscalar approach to the production of geographic context. *International Journal of Geographical Information Science*, 33(6), 1270–1288.

Sapountzi, A., and Psannis, K. (2020). Big data preprocessing: an application on online social networks, in H. Arabnia, K. Daimi, R. Stahlbock, C. Soviany, L. Heilig, and K. Brüssau (Eds.), *Principles of data science*. Cham, Switzerland: Springer Nature, 49–78.

Uncertainty in the effects of the modifiable areal unit problem under different levels of spatial autocorrelation: a simulation study

Sang-Il Lee ⓘ, Monghyeon Lee ⓘ, Yongwan Chun ⓘ and Daniel A. Griffith ⓘ

ABSTRACT

The objective of this paper is to investigate uncertainties surrounding relationships between spatial autocorrelation (SA) and the modifiable areal unit problem (MAUP) with an extensive simulation experiment. Especially, this paper aims to explore how differently the MAUP behaves for the level of SA focusing on how the initial level of SA at the finest spatial scale makes a significant difference to the MAUP effects on the sample statistics such as means, variances, and Moran coefficients (MCs). The simulation experiment utilizes a random spatial aggregation (RSA) procedure and adopts Moran spatial eigenvectors to simulate different SA levels. The main findings are as follows. First, there are no substantive MAUP effects for means. However, the initial level of SA plays a role for the zoning effect, especially when extreme positive SA is present. Second, there is a clear and strong scale effect for the variances. However, the initial SA level plays a non-negligible role in how this scale effect deploys. Third, the initial SA level plays a crucial role in the nature and extent of the MAUP effects on MCs. A regression analysis confirms that the initial SA level makes a substantial difference to the variability of the MAUP effects.

Introduction

The modifiable areal unit problem (MAUP) refers to the arbitrary nature of areal units used in many spatial analyses, as well as the dependency of resulting statistical properties upon the spatial configuration of these areal units (Wong 2009a, Wong 2009b). A configuration of areal units employed in a study is *modifiable*, or more accurately *substitutable*, because many alternative surface partitionings exist, which are actually available and/or theoretically viable. Although, in some situations, a specific areal unit configuration is essential because of data availability only with that particular configuration, in other situations, one configuration can be preferred to others. In addition, researchers may constitute a new configuration by aggregating pre-existing lower-level areal units (i.e. smaller polygons). In any of the preceding cases, the arbitrary nature of areal units is unavoidable such that no undeniable justification is possible regarding whether or not one spatial configuration is optimal for revealing an underlying spatial process of a phenomenon under investigation. The twin analytical results aspect of the MAUP, their dependency upon or *sensitivity* to a spatial configuration (Fotheringham and Wong 1991, 1025), is more fundamental; a different spatial configuration often yields significantly different statistical results. This *uncertainty* or *instability* of analytical results (Fotheringham and Wong 1991, Manley 2014) implies that no conclusive statistical statement is possible in the field of spatial analysis, especially when areal data are used.

The vast majority of MAUP studies have been dedicated to exploring and analyzing how significant the effects of the MAUP are, and in which ways they have an impact on statistical results, not only for such basic descriptive statistics as means, variances, and correlation coefficients but also for more sophisticated statistical techniques, such as multiple regression and other types of spatial data analyses (e.g. Arbia and Petrarca, 2011, Fotheringham and Wong 1991, Amrhein 1995, Amrhein and Reynolds 1996, Wong *et al.* 1999, Flowerdew *et al.* 2001, Dark and Bram 2007, Arbia and Petrarca 2011). Even though a considerable amount of literature has accumulated especially since the mid-1990s, our knowledge regarding both diagnosis and prognosis of the MAUP is still limited. Indeed, an observation made about 35 years ago by one of the earlier pioneers in the MAUP research is still valid (Openshaw, 1984, 6): 'the MAUP is today one of the most important unresolved problems left in spatial analysis.' This sentiment is well echoed by a recent review of the MAUP (Manley 2014, 1158); 'we have neither a full and detailed understanding of the problem nor the underlying causes.' Hence, more effort is necessary to develop a research framework to obtain more comprehensive, and possibly more generalizable, results about how the MAUP effects behave.

Spatial autocorrelation (SA) is known to be a primary source of the MAUP (Openshaw and Taylor 1979, Arbia 1989, Fotheringham and Wong 1991, Wong 1996), and efforts to discover a relationship between the level of aggregation (AG) and the level of SA have been made (Cliff and Ord 1981, Chou 1991, 1995, Qi and Wu 1996, Griffith *et al.* 2003). Also, an impact of spatial aggregation on SA has been well investigated in geostatistics (e.g. Journel and Huijbregts 1978). Especially, the effect of regularization on a variogram (that is, how the overall structure of SA changes with spatial aggregation) is well explored in the context of change of support. Recent studies, including Kyriakidis (2004), Kyriakidis and Yoo (2005), and Yoo *et al.* (2010), explore impacts of spatial aggregation in area-to-point spatial interpolation, focusing more on scale effects. However, much of the interplay between these two concepts, once referred to as 'two very stubborn but pervasive problems in statistical analysis of spatial data' (Wong 2009a, 120), still remains unknown. That is, SA is a source of uncertainty in the MAUP effects that make it difficult to derive a generalizable behavior for the MAUP. In addition, despite a consensus that a well-designed simulation is essential to a solid research framework to evaluate the effects of the MAUP in a statistical analysis (Green and Flowerdew 1996, 43), methodological advances have been meager. A better simulation framework may require a well-founded random aggregation procedure (e.g. Flowerdew *et al.* 2001), which is equipped with a reliable and efficient algorithm for aggregating areal units for different levels of AG. It also should have a conceptually sound evaluation scheme furnishing a simultaneous assessment of both the scale and zoning effects on statistical properties.

The objective of this paper is to investigate uncertainty surrounding relationships between SA and the MAUP with an extensive simulation experiment. Although the literature shows that they have an impact on each other, it is still uncertain how they affect each other. For instance, Fotheringham and Wong (1991) show how the MAUP can behave differently with four census variables that have various levels of SA, but it is limited to only the empirical variables and is not enough to explore a wide spectrum of uncertainty. Hence, this paper aims to explore how differently the MAUP behaves across levels of SA. Specifically, the investigation focuses on whether the initial level of SA at the finest spatial scale makes a substantial difference to the MAUP effects, the scale effect arising from the level of aggregation, and/or the zoning effect arising from the variety of zonations at the same AG level. That is, the level of SA at the finest resolution is considered as a factor that increases uncertainty of the MAUP effects. The initial level of SA as a potential factor on the MAUP is visualized and examined with a regression analysis using the outcome of the simulation experiments, an assessment not appearing in the literature. The impacts on three univariate summary statistics are focused on: i.e. the mean, variance, and Moran coefficient (MC). In the simulation experiment, a random spatial aggregation (RSA) procedure was devised and utilized to generate random zonations by aggregating smaller areal units.

Spatial autocorrelation and the MAUP

Two major MAUP effects exist: scale and zoning (also referred to as zonation or aggregation). Assuming that the overall MAUP effects occur in a *spatial aggregation process* (the same as a *spatial partitioning process* in a theoretical sense) whereby 'a larger number of smaller areal units are grouped into a smaller number of larger areal units' (Amrhein 1995, 105), the two sub-effects are jointly responsible for the complete process. The scale effect occurs because of differences in the number of areal units into which a study region has been partitioned. In contrast, the zoning effect occurs exclusively because of differences in how lower-level areal units are grouped into a particular number of higher-level areal units. The importance of SA in MAUP studies, or the interplay of these two concepts, is twofold. First, SA is a primary source of the MAUP. Second, SA itself is subject to the MAUP effects.

Regarding the first aspect, Fotheringham and Wong (1991) and Wong (1996) explicitly point out a direct link between the two, which was suggested earlier by Openshaw and Taylor (1979). A *smoothing process* occurs when spatial aggregation proceeds, and is responsible for a tendency of reduced variance and correlation. This explanation seems to apply at least to the scale effect (Green and Flowerdew 1996, Wong 1996). As adjacent areal units are aggregated to constitute a larger areal unit, their peculiarities or heterogeneity are expected to be reduced, thus resulting in a reduction in variance and correlation coefficients, assuming a relatively stable covariance (Fotheringham and Wong 1991). Furthermore, Wong (1996) argues that the degree of susceptibility to the MAUP effects could vary from one variable to another because they contain different levels of SA, which may explain why succinct results from MAUP studies dealing with statistical situations involving multiple variables are more difficult to obtain.

The zoning effect, even at some given spatial scale, also can lead to uncertainty or instability in a spatial data analysis. As Openshaw (1984) points out, the zoning effect may be greater than the scale effect. Lee (2001) proposes a spatial smoothing scale, which is subsequently named S, as an alternative univariate SA measure (Lee 2004, 2009, 2017). This particular measure is based on the concept that the SA level of a geographic variable is directly associated with the amount of variance reduction attributable to transforming a variable to a spatial lag vector or a spatial moving average vector. For example, while the least variance reduction occurs when a variable has extreme positive SA, the most variance reduction occurs when a variable has extreme negative SA (see Figure 2 in Lee 2001). This correspondence implies that the zoning effect is closely related to local SA contexts. That is, if a set of neighboring areal units with strong local positive SA are aggregated into a larger areal unit, the aggregation makes no or little contribution to variance reduction. In contrast, an aggregation of a set of neighboring areal units with strong negative SA contributes to a larger variance reduction.

Regarding the second aspect, SA measures such as the MC are themselves subject to the MAUP effects. Cliff and Ord (1981) report that there is a negative relation between AG and SA levels, showing that the larger the size of areal units, the smaller the MC value tends to be. Similarly, Chou (1991, 1995) discusses a possibly generalizable relationship (a log-linear one) between map resolution and the MC, and Qi and Wu (1996) also report the same observation form their analyses with landscape pattern data. Similarly, Griffith et al. (2003) show that the SA level measured by the MC decreases as the spatial resolution of areal units gets coarser, from block groups through counties to states. Despite these observations of salient trends, little literature provides substantive explanations for this situation.

The aforementioned smoothing process (i.e. variance reduction) proposition may provide a possible primitive explanation for a relationship between AG and SA levels. The variance reduction decreases the MC denominator value and, subsequently, results in an increase in the MC value. Simultaneously, variance reduction tends to decrease the spatial covariance value in the MC numerator, and subsequently also decreases the MC value. Thus, the MC usually shrinks toward zero when the reduction of its numerator is larger than the reduction of its denominator. Although this explanation is more directly related to the scale effect, the same argument can be used for the zoning effect. That is, different zonations trigger a different local SA heterogeneity that may lead to differences in reduction of the numerator and the denominator values.

In addition, a worthwhile investigation would be to examine whether or not the negative relationship between AG and SA levels can behave differently based on an initial SA level. In other words, it is unclear whether or not a variable with a higher MC value is more sensitive to the MAUP effects than one with a lower value when they are aggregated. Although previous work (Chou 1991) claims that an initial SA level plays a non-negligible role, this contention largely remains under-investigated. Specifically, how and how much an initial SA level influences the MAUP effects, scale and/or zoning, has received little attention.

Research design

This paper utilizes experimental simulations to elucidate the nature and extent of the influence of the initial SA level on the MAUP effects. Changes in univariate statistical values as well as SA measures of variables are monitored along with different AG levels and different zonations of spatial units. An RSA procedure was utilized to construct a coarse (aggregated) spatial tessellations from fine spatial units: that is, aggregating small polygons into fewer larger polygons at a coarser spatial resolution. The detailed procedure of the RSA is as follows: suppose that n original areal units are to be aggregated into m target areal units then

(1) A set of seed units (m) is randomly selected from the original areal units (n).
(2) In the first round, each seed unit, in a random order, annexes a randomly chosen neighboring unit to construct a first-round zone. When a seed does not have any available neighboring units, the seed unit itself becomes a final zone for the target tessellation. For other first-round zones, the aggregation proceeds to the second round.
(3) In the second round, each of the first-round zones, in a random order, annexes one of the remaining units (that are neither a seed unit nor an annexed unit in the first round) that are contiguous to any of its participating subunits to construct a second-round zone. If no neighboring unit is available for annexation for a first-round zone because all neighboring units already are annexed into other first-round zones, it becomes a final zone for the target tessellation. For other second-round zones, the aggregation proceeds to the third round.
(4) This procedure continues to move to additional rounds until all areal units are annexed to one of m target units.

The RSA imposes two restrictions on the random selection process for neighbors. First, 'no aggregation' is given as an option such that, for instance, when only one neighboring unit is available for annexation, the probability of the neighboring unit being selected is not 1 but 1/2. This restriction should better ensure the nature of randomness for a spatial aggregation. Second, from the second round on, a neighboring unit that is contiguous to more subunits of a zone has a higher chance of being selected. For instance, when a zone with two subunits has two candidates' neighboring units for annexation, one neighbor that is adjacent to both subunits has a higher chance of being selected than the other that is adjacent to one of the two subunits; the former has a selection probability that is twice that of the latter. This restriction helps avoid contorted zones (e.g. gerrymandering-type zones) and enhances the compactness of resulting zones (Flowerdew *et al.* 2001).

The aggregation process begins with a regular tessellation of 1,024 squares (32-by-32), which is the finest spatial resolution in this simulation (Figure 1(a)): see Boots and Tiefelsdorf (2000) for a more detailed description. These squares are randomly aggregated into 10 different AG levels (i.e. coarser resolutions) with the rook-type spatial neighboring structure: AG1 (896), AG2 (768), AG3 (640), AG4 (512), AG5 (384), AG6 (256), AG7 (128), AG8 (64), AG9 (32), and AG10 (16). Note that the numbers in the parentheses indicate the number of areal units for a target (aggregated) tessellation. The regular square tessellation was chosen over other types of regular tessellations (e.g. hexagons as seen in Figure 1(b)) because a regular square tessellation, with a rook adjacency definition, allows a full and symmetric range of SA from highly positive to equally highly negative (Boots and Tiefelsdorf 2000). For each AG level, 1,000 different sets of zonations are generated. The two different sources of variability for the MAUP are incorporated in this simulation design: variability owing to the different AG levels (i. e. the scale effect), and variability owing to the different zonations (i.e. the zoning effect).

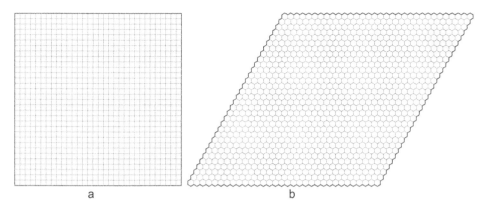

Figure 1. Two regular tessellations ($n = 1,024$): (a) Squares and (b) Hexagons.

Spatially autocorrelated random variables were constructed with Moran eigenvectors, which are fundamental components of the eigenvector spatial filtering methodology (Griffith 1996, 2000, 2003, Tiefelsdorf and Griffith 2007). These eigenvectors provide a set of orthogonal and uncorrelated vectors that portray distinct SA patterns. Importantly, their corresponding eigenvalues essentially are MC values for them (Griffith 1996). Hence, they provide a set of numerical values covering a full range of possible SA for the spatial tessellation employed here, from extreme positive to extreme negative SA. To explore different SA levels, nine spatial eigenvectors representing various SA levels were selected from a total of 1,024 eigenvectors, which are extracted from a transformed spatial weights matrix for the 1,024 squares forming the tessellation: SA1 (EV1, MC = 1.0206), SA2 (EV91, MC = 0.7471), SA3 (EV192, MC = 0.5037), SA4 (EV317, MC = 0.2487), SA5 (EV529, MC = −0.0070), SA6 (EV707, MC = −0.2487), SA7 (EV832, MC = −0.5037), SA8 (EV933, MC = −0.7471), and SA9 (EV1024, MC = −1.0276). These nine eigenvectors were selected using approximately equal MC spacing (about 0.25), from extreme positive to extreme negative SA, covering the entire feasible SA range.

These eigenvectors have the same mean of zero, and the same variance of $1/1023 \approx 0.0010$, except for a few eigenvectors in the middle of the spectrum,[1] which ensures a controlled initial univariate condition. For each aggregated zonation, three summary statistics (i.e. the mean, variance, and MC) are computed and recorded, which results in 90,000 means, variances, and MCs (i.e. 10 AG levels times 1,000 zonations times nine SA levels). The behavior of the three summary statistics is inspected to explore the way and the extent to which the initial SA level contributes to the variability of the MAUP effects, scale and/or zoning. Finally, regression is employed to numerically determine how much the summary statistics are influenced by the effects of the MAUP at a given initial level of SA and at a given AG level.

For a sensitivity analysis, the same experiment is conducted for a regular hexagon tessellation to check any potential impacts of a spatial tessellation type (Figure 1(b)). Empirical spatial tessellations (e.g. census units) tend to be somewhere between a hexagon and a square tessellation. However, the spectrum of the eigenvalues extracted from a spatial hexagon tessellation is not symmetric, unlike the one from a square tessellation with the rook type spatial neighboring structure (Boots and Tiefelsdorf 2000). Some informative results are presented in the Appendix to this paper.

Results

This section presents the results of the simulation experiments. The first discussion summarizes findings about the variability of the MAUP effects on the following three summary statistics: the mean, variance, and MC. The second discussion presents the results of regression analyses that were conducted to explore a potential systematic relationship between the MAUP effects and AG and initial SA levels.

The MAUP effects on the means, variances, and MCs

For each summary statistic, nine sets of 10 boxplots are presented: each boxplot visualizes the expectation and dispersion for 1,000 zonations for each of the 10 AG levels; sets of these 10 boxplots are drawn for each of the nine SA levels. Figure 2 shows the MAUP effects on the means. Except for SA1, the graphs have a similar pattern; the means are consistently around zero across all AG levels (no or little scale effect) and are compactly dispersed in a narrow range for each AG level (weak zoning effect). This outcome tends to support the common belief that the MAUP does not have a significant impact on means (Amrhein 1995). However, the result of SA1 does not support this contention. Its pattern is quite exceptional; the dispersion steadily increases as the AG level increases, which implies that an extremely high SA level seems to be largely susceptible to the zoning effect, especially at high AG levels. A global SA pattern of SA1, which has two spatial clusters of large and small values, may lead to some aggregated units with extremely large or small means. An inspection of all the 1,024 SA levels (i.e. eigenvectors) reveals that this high variability is observed for the first 10 eigenvectors. The implication is that there is no (or at least a very weak) scale effect, although an initial SA level plays a role in the deployment of the zoning effect on means when extreme positive SA is present.

Figure 3 shows the MAUP effects on the variances. Except for the SA1 case, the graphs show a similar trend that as the AG level increases, variance systematically decreases, flattening out around zero, which may indicate a strong scale effect. The range of the variances is relatively stable across the AG levels, which may indicate a weak and constant zoning effect. This result tends to comply with the variance reduction reported in the literature (Fotheringham and Wong 1991, Wong 1996). However, the initial SA level plays a non-negligible role. First, as the initial SA level gets lower, the flattening pattern appears at a lower AG level. Specifically, it occurs around AG9 for SA3–SA4, AG8 for SA5–SA7, and AG7 for SA8–SA9. Second, the result of SA1 is different than that for the other cases. The variances for the SA1 results do not decrease as sharply as those for the other SA levels, and never flatten out. In addition, the dispersion of the variances increases as the AG level increases, implying that an extremely high SA level seems to be more susceptible to the zoning effect at higher AG levels. These results for SA1 may originate from its spatial pattern with two clusters of large and small values again, which leads to considerable variance remaining in its aggregated values. In sum, the scale effect is larger than the zoning effect, and an initial SA level plays a non-negligible role in the deployment of both effects.

Figure 4 shows the MAUP effects on the observed MC values. The distinct patterns of the nine graphs indicate that the initial SA level plays a pronounced role in the MAUP

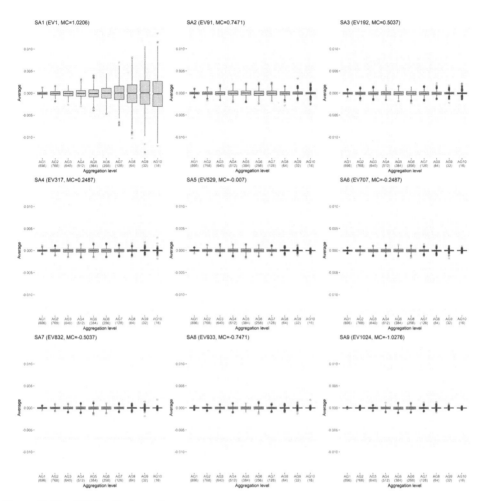

Figure 2. The MAUP effects on the means.

effects on MCs. In detail, four points are noteworthy. First, when an initial SA level is high, as the AG level increases, the MC decreases constantly until it converges to around −0.25. This convergence occurs early when the initial SA is low. In contrast, when the initial SA is negative, the MC increases and converges to around −0.25. Second, after the MC converges, it tends to oscillate, which can be easily observed in the results of SA6. Third, variability tends to increase as the AG level increases, which is conspicuous in all of the nine graphs. Fourth, the resulting patterns are not symmetric between positive and negative SA. For example, the MC values for SA1 decrease slowly for AG1–AG6, and then rapidly decrease beyond AG7. In contrast, the MC values for SA9 change more rapidly for AG1–AG4. In addition, the zoning effect is much larger in SA9; its boxplots show larger variances than those displayed by SA1. However, the results of SA5 and SA7 show a symmetric pattern, around the convergence point (−0.25), which is roughly the initial SA level of SA6. These results provide counterexamples to a general statement in the literature that the MC systematically decreases as the AG level increases (e.g. Chou

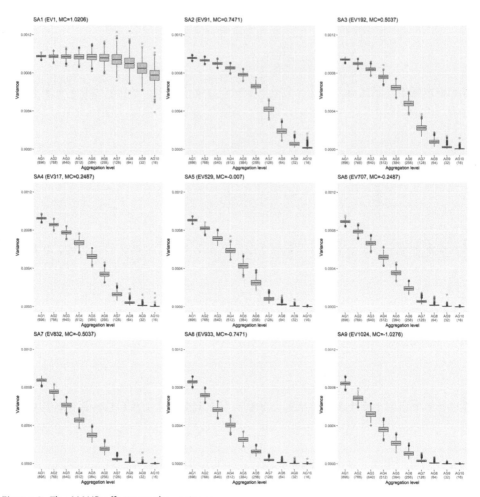

Figure 3. The MAUP effects on the variances.

1991, 1995, Qi and Wu 1996). It is only partially true for some AG levels (i.e. AG1–AG6) coupled with higher positive SA levels (i.e. SA2–SA4).

Standardized MCs are also examined here, to control effects of the different expected values and variances with different numbers of observations (Figure 5). Unlike in Figure 4, the prominence of the zoning effect at high AG levels disappears. That is, dispersions are relatively constant across all AG levels. However, the scale effect is more clearly seen. For higher positive SA levels, a standardized MC systematically decreases, flattening out around zero, and for highly negative SA levels, a standardized MC steadily increases, flattening out around zero. There is little or no scale effect exhibited in moderate to no SA levels.

Formulating the MAUP effects

Figure 6 presents a pair of line graphs for the means: one for averages (the scale effect), and the other for variances (the zoning effect) of the summary statistics from the 1,000

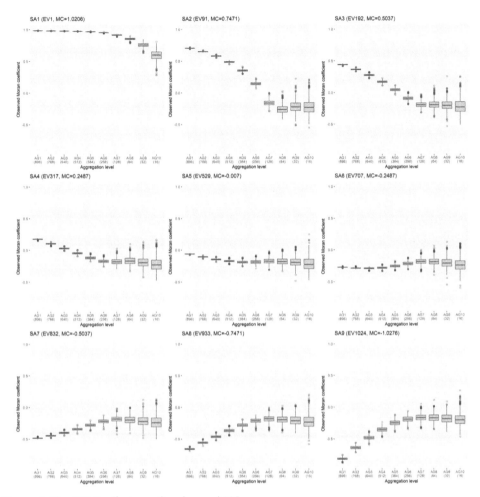

Figure 4. The MAUP effects on the observed MC.

zonations for each combination of AG levels and initial SA levels. That is, each dot in Figure 6 represents the average (a) and the variance (b) of a corresponding boxplot in Figure 2; a total of 90 means and 90 variances are represented in Figure 6. Similarly, Figures 7–9 present the same type of information, respectively, for the variances, observed MCs, and standardized MCs.

Figure 6(a,b) indicates an absence of the scale effect and the zoning effect, if SA1 is set aside. That is, the means are around zero in Figure 6(a), and the variances are close to zero in Figure 6(b). Figure 7(a) supports the variance reduction proposition across all SA levels. It also shows that variance more rapidly decreases when an initial SA level is low. Figure 7(b) shows that the zoning effect occurs differently based on positive and negative SA, excluding SA1. While positive SA results have large variances at high AG levels (i.e. AG6–AG8), negative SA cases have large variances at low AG levels (i.e. AG3–AG4). Figure 8(a) portrays the relationship between the AG and the SA levels. It shows that their relationship is opposite between positive and negative SA for the initial SA level. However, all lines, except SA1, converge as the AG level gets larger. The extent of

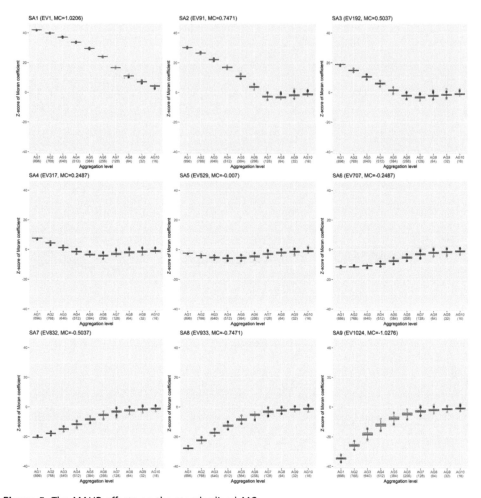

Figure 5. The MAUP effects on the standardized MC.

MC value changes from the initial MC at each AG level indicate that the scale effect is stronger with a higher SA level, regardless of sign, and that the symmetry between positive and negative SA is not obvious or weaker than one might expect. Figure 8(b) shows that the AG level is the predominant factor explaining the zoning effect for the MCs; that is, the MC varies more across 1,000 zonations when the AG level is high. Although Figure 9(a) conveys almost the same information as Figure 8(a), Figure 9(b) presents a potential impact of the sign of the initial SA level (i.e. positive and negative SA) in the zoning effect for the standardized MCs.

A regression analysis was conducted between the MAUP effects and AG and initial SA levels to explore a systematic relationship between them. Such a regression analysis helps to formulate the nature and extent of MAUP effects when aggregation occurs for an AG level and a SA level. Amrhein and Reynolds (1996, 1997) explored this relationship in a similar way; their dependent variable was derived (i.e. not observed), which was believed to capture the overall aggregation effect. The main goal of this regression analysis was to examine whether or not the initial SA level leads to a significant

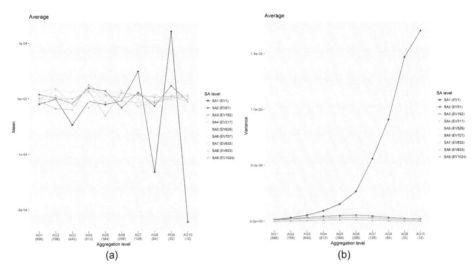

Figure 6. The scale effect (a) and the zoning effect (b) on the means by the initial level of spatial autocorrelation.

difference in the MAUP effects. Two regression models were fitted with the data points in Figures 6–9: one for the scale effect with the average data points in Figures 6(a)–9(a), and the other for the zoning effect with the variance data points in Figures 6(b)–9(b). The two major independent variables are the AG level (*AG*; the number of areal units) and the initial SA level (*SA*; the observed MCs at the finest level). A dummy variable (*D_SA*) was used to control the influence of positive or negative SA in the initial SA level: 1 for positive SA and 0 for negative SA. In addition, two statistical interaction variables, *AG*D_SA* and *SA*D_SA*, also were included to examine whether or not the sign of *SA* leads to a different relation for the two major independent variables

Table 1 summarizes the regression results. For the means, there are two notable findings. First, no scale effect is observed. Second, the zoning effect is stronger when the SA level is positive (the *SA*D_SA* is significant). Also, the *AG*D_SA* is significant at the 1% level, although potentially affected by the results of SA1, which have increasing variances as the AG level increases, unlike the other cases. Indeed, a supplementary regression analysis without the SA1 results shows that the *AG*D_SA* is not significant, but *SA* is significant at the 5% level.[2] This supplemental regression result may indicate that the initial SA level is significantly associated with the zoning effect (*SA* is significant at the 5% level with a positive sign). That is, as the SA level increases, the zoning effect gets larger.

For variance, there are three notable outcomes. First, the scale effect is obvious, with significant coefficients for *AG* and *SA*. The AG level is a dominant factor (*t*-value of 18.077), indicating that variance decreases as the AG level increases (note that because this variable is the number of areal units, its positive sign is indicative of a negative relationship between the AG level and the variance). Second, the initial SA level plays a significant but secondary role. With an AG level being constant, the variance gets larger as the initial SA level becomes higher (*t*-value of 2.103 for *SA*). This relationship is stronger for positive SA (*t*-value of 6.283 for *SA*D_SA*). Third, for the zoning effect, the

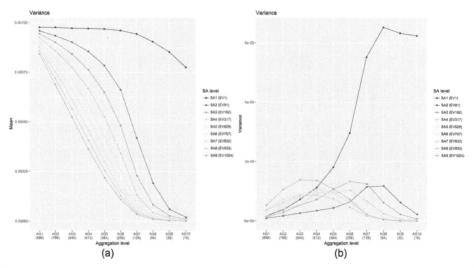

Figure 7. The scale effect (a) and the zoning effect (b) on the variances by the initial level of spatial autocorrelation.

two interaction terms are highly significant, suggesting the importance of the differences between positive and negative SA. The significance of $AG*D\_SA$ with a negative sign indicates that when SA is positive, the relationship is reversed such that the variance increases as the AG level increases, which can be seen by the change of the AG coefficient from 0.0074 for negative SA, to −0.0022 for positive SA levels. The significance of the $SA*D\_SA$, however, seems to be affected by the exceptional SA1 results because it is not significant in the supplemental regression that excludes the SA1 results.

For the MCs, there are three notable outcomes. First, the AG level is a substantial factor, with a significant coefficient at the 1% level (−0.0030). However, when $AG*D\_SA$, which is significant with a coefficient of 0.0097, is considered, its relationship gets reversed for positive SA. Second, the initial SA level is also an important factor when it is referenced by its sign. That is, the MC values are higher for higher initial SA levels, and this relationship is stronger for positive SA ($SA*D\_SA$ is extremely significant with a t-value of 9.975). Third, the AG level is the only significant factor for the zoning effect, but the model explains almost half of the variability in the variance. The results of standardized MCs show the same associations for the scale effect as for the observed MCs. In contrast, both the AG level and the initial SA level are significant for the zoning effect; furthermore, their relationships are significantly different for positive and negative SA cases (D_SA is significant at the 1% level).

Conclusions

This paper investigates uncertainty in the MAUP effects focusing on SA using simulation experiments. While most studies that have investigated an association between SA and the MAUP utilize a limited set of empirical variables (e.g. Fotheringham and Wong 1991, Amrhein and Reynolds 1997), this paper examines a wide range of SA in the experiments with Moran spatial eigenvectors. Important findings of this paper include the following two outcomes.

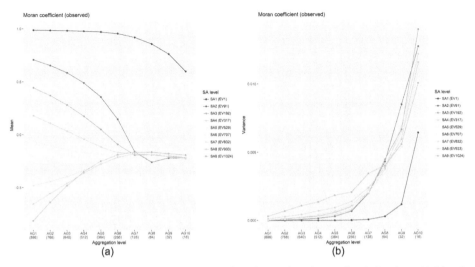

Figure 8. The scale effect (a) and the zoning effect (b) on the observed MCs by the initial level of spatial autocorrelation.

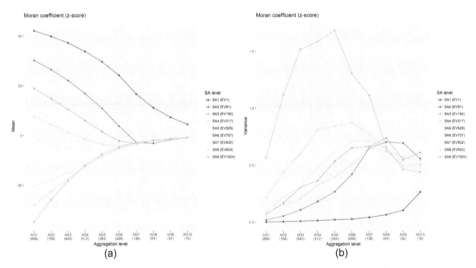

Figure 9. The scale effect (a) and the zoning effect (b) on the standardized MCs by the initial level of spatial autocorrelation.

First, the simulation results support the contention that the initial SA level makes a marked difference in the variability of the MAUP effects, and increases uncertainty in the MAUP effects. That is, the nature and extent of the MAUP effects substantially vary due to the initial SA level of a variable as spatial aggregation proceeds. This outcome means that the initial SA level is a key factor for a plausible answer to the following question: by how much and how is SA susceptible to the MAUP effects? Second, this paper shows that the scale effect in MCs occurs with various tendencies (see Figure 8(a)), and the zoning effect is severe (see Figure 8(b)). The convolution of these effects makes uncovering a systematic MAUP effect difficult. Notably,

Table 1. Regression results for the square tessellations.

	Means		Variances		Observed MCs		Standardized MCs	
	Scale effect (Mean)	Zoning effect (Variance)	Scale effect (Mean)	Zoning effect (Variance)	Scale effect (Mean)	Zoning effect (Variance)	Scale effect (Mean)	Zoning effect (Variance)
Intercept	-1.4974	10.8215	-0.0436	2.2668	-0.7058	5.3124	3.3271	0.3985
	(-0.152)	(0.174)	(-0.012)	(0.790)	(-1.545)	(6.815***)	(2.334*)	(5.704***)
AG	0.0026	0.0028	0.1029	0.0074	-0.0030	-0.0080	-0.0204	-0.0003
	(0.168)	(0.029)	(18.077***)	(1.665)	(-4.212***)	(-6.594***)	(-9.209***)	(-2.785**)
SA	-2.9507	6.6093	10.1751	-0.4059	1.8017	-0.6181	7.3080	-0.5993
	(-0.228)	(0.081)	(2.103*)	(-0.107)	(2.991**)	(-0.601)	(3.888***)	(-6.504***)
D_SA	9.5024	-98.5848	-10.5736	-1.8205	-6.7480	1.3144	-24.2238	0.4178
	(0.553)	(-0.910)	(-1.650)	(-0.363)	(-8.458***)	(0.965)	(-9.730***)	(3.423***)
AG*D_SA	0.0107	-0.4063	-0.0161	-0.0296	0.0097	0.0002	0.0491	-0.0003
	(0.469)	(-2.815**)	(-1.889)	(-4.428***)	(9.151***)	(0.089)	(14.799***)	(-1.598)
SA*D_SA	-26.6841	606.0092	52.3857	31.8907	10.3543	-1.6368	24.4124	0.1462
	(-1.194)	(4.298***)	(6.283***)	(4.888***)	(9.975***)	(-0.924)	(7.536***)	(0.920)
R^2	0.0482	0.3855	0.8952	0.4513	0.8743	0.4881	0.8903	0.5733

Note 1. Numbers in parentheses are t-values for the regression coefficients. Significance codes: ***: 0.001; **: 0.01; *:0.05

2. The dependent variables are multiplied by a constant from the smallest 10 to the largest 10^{10} since their data scale is much smaller than the independent variable. Hence the coefficients for a given variable do not reflect the magnitude of differences across the different summary statistics (i.e. mean, variance, observed MC, and standardized MC).

this paper is the first report that presents a regression analysis result to confirm the impacts of potential factors on the MAUP effects. In addition, this paper effectively visualizes the MAUP effects with boxplots that graphically illustrate the impact of initial SA levels. Since datasets at a fine spatial resolution increasingly have been available (e.g. remotely sensed images), and often are processed for data reduction (i.e. aggregation), understanding how the MAUP effects change in conjunction with initial SA levels is very important.

The findings of this paper provide encouraging evidence for further investigating the MAUP effects in multivariate situations in future research. That is, the identification of the initial SA level as a key factor and the development of the RSA procedure can furnish a sound foundation to investigate the MAUP effects for multivariate statistics such as regression (e.g. Amrhein and Reynolds 1997), which is still under investigated. Fotheringham and Wong (1991) report that there is little connection between the MAUP effects and the SA degree in multiple regression. Similarly, Flowerdew *et al.* (2001) conclude that the SA level is important for correlation coefficients, but not for regression coefficients. Although some efforts have been made to decipher this ambiguity (among others, Arbia 1989, Green and Flowerdew 1996, Flowerdew *et al.* 2001, Manley *et al.* 2006), no definitive answers have been provided.

This research also can be extended to examine MAUP effects in correlation coefficients. Considerable efforts addressing this topic can be found in the literature (e.g. Openshaw and Taylor 1979, Openshaw, 1984; Fotheringham and Wong 1991, Amrhein 1994, 1995); one common finding is that the correlation level increases as the AG level increases. However, many other aspects have yet to be investigated. First, the impacts of the initial level of correlation on the MAUP effect are under-investigated. The MAUP effect drawn from an experimental simulation with highly positive correlation might be substantially different from one with no correlation, or highly negative correlation. Second, the initial level of bivariate SA also may have a considerable impact on the variability of MAUP effects. This task may require a proper measure for the level of bivariate SA among possible options, including bivariate MC and Lee's L statistics (Lee 2001, 2004), particularly L^* (Lee 2017). In addition, a much more complicated simulation framework is needed to employ both the initial level of correlation and the initial level of bivariate SA, which is beyond the scope of this research.

Notes

1. These eigenvectors with an MC of zero have non-zero means and less variance because they are associated with an eigenvalue with multiplicity 32.
2. Results of the supplemental regression are not presented in this paper because the variable significances are the same at the 5% level except highlighted variables in this text. These variables are, for the zoning effect, (1) *SA & SA*D_SA* for the mean, (2) *AG & SA*D_SA* for the variance, and (3) the two interaction terms for the standardized MCs.

Disclosure statement

No potential conflict of interest was reported by the authors.

ORCID

Sang-Il Lee ⓘ http://orcid.org/0000-0002-5342-9930
Monghyeon Lee ⓘ http://orcid.org/0000-0002-5778-5967
Yongwan Chun ⓘ http://orcid.org/0000-0002-4957-1379
Daniel A. Griffith ⓘ http://orcid.org/0000-0001-5125-6450

References

Arbia, G. and Petrarca, F., 2011. Effects of MAUP on spatial econometric models. *Letters in Spatial and Resource Sciences*, 4, 173–185. doi:10.1007/s12076-011-0065-9

Amrhein, C.G., 1994. Testing the use of a hybrid regionalization scheme for confidential tax-filer data. *Canadian Journal of Regional Science*, 17, 259–274.

Amrhein, C.G., 1995. Searching for the elusive aggregation effect: evidence from statistical simulations. *Environment and Planning A*, 27, 105–119. doi:10.1068/a270105

Amrhein, C.G. and Reynolds, H., 1996. Using spatial statistics to assess aggregation effects. *Geographical Systems*, 2, 143–158.

Amrhein, C.G. and Reynolds, H., 1997. Using the Getis statistic to explore aggregation effects in metropolitan Toronto census data. *The Canadian Geographer*, 41, 137–149. doi:10.1111/j.1541-0064.1997.tb01154.x

Arbia, G., 1989. *Spatial data configuration in statistical analysis*. Norwell: Kluwer Academic Publishers.

Boots, B. and Tiefelsdorf, M., 2000. Global and local spatial autocorrelation in bounded regular tessellations. *Journal of Geographical Systems*, 2, 319–348. doi:10.1007/PL00011461

Chou, Y.H., 1991. Map resolution and spatial autocorrelation. *Geographical Analysis*, 23, 228–246.

Chou, Y.H., 1995. Spatial pattern and spatial autocorrelation. In: A.U. Frank and W. Kuhn, eds. *Spatial information theory: a theoretical basis for GIS, Proceedings of international conference COSIT 95*, Semmering, Austria. Berlin: Springer, 365–376.

Cliff, A.D. and Ord, J.K., 1981. *Spatial processes: models & applications*. London: Pion.

Dark, S.J. and Bram, D., 2007. The modifiable areal unit problem (MAUP) in physical geography. *Progress in Physical Geography*, 31 (5), 471–479. doi:10.1177/0309133307083294.

Flowerdew, R., Geddes, A., and Green, M., 2001. Behaviour of regression models under random aggregation. *In*: N. Tate and P.M. Atkinson, eds. *Modelling scale in geographical information science*. Hoboken: John Wiley & Sons, 89–104.

Fotheringham, A.S. and Wong, D.W.S., 1991. The modifiable areal unit problem in multivariate statistical analysis. *Environment and Planning A*, 23, 1025–1044. doi:10.1068/a231025

Green, M. and Flowerdew, R., 1996. New evidence on the modifiable areal unit problem. *In*: P. Longley and M. Batty, eds. *Spatial analysis: modelling in a GIS environment*. Cambridge: GeoInformation International, 41–54.

Griffith, D.A., 1996. Spatial autocorrelation and eigenfunctions of the geographic weights matrix accompanying geo-reference data. *The Canadian Geographers*, 40, 351–367. doi:10.1111/j.1541-0064.1996.tb00462.x

Griffith, D.A., 2000. A linear regression solution to the spatial autocorrelation problem. *Journal of Geographical Systems*, 2, 141–367. doi:10.1007/PL00011451

Griffith, D.A., 2003. *Spatial autocorrelation and spatial filtering: gaining understanding through theory and scientific visualization*. Berlin: Springer.

Griffith, D.A., Wong, D.W.S., and Whitfield, T., 2003. Exploring relationships between the global and regional measures of spatial autocorrelation. *Journal of Regional Science*, 43, 683–710. doi:10.1111/j.0022-4146.2003.00316.x

Journel, A.G. and Huijbregts, C.J., 1978. *Mining geostatistics*. London: Academic press.

Kyriakidis, P.C., 2004. A geostatistical framework for area-to-point spatial interpolation. *Geographical Analysis*, 36 (3), 259–289. doi:10.1111/j.1538-4632.2004.tb01135.x.

Kyriakidis, P.C. and Yoo, E.H., 2005. Geostatistical prediction and simulation of point values from areal data. *Geographical Analysis*, 37 (2), 124–151. doi:10.1111/j.1538-4632.2005.00633.x.

Lee, S.-I., 2001. Developing a bivariate spatial association measure: an integration of Pearson's *r* and MC. *Journal of Geographical Systems*, 3, 369–385. doi:10.1007/s101090100064

Lee, S.-I., 2004. A generalized significance testing method for global measures of spatial association: an extension of the Mantel test. *Environment and Planning A*, 36, 1687–1703. doi:10.1068/a34143

Lee, S.-I., 2009. A generalized randomization approach to local measures of spatial association. *Geographical Analysis*, 41, 221–248. doi:10.1111/j.1538-4632.2009.00749.x

Lee, S.-I., 2017. Correlation and spatial autocorrelation. *In*: S. Shekhar, H. Xiona, and X. Zhou, eds.. *Encyclopedia of GIS*, Vol. 1, 2nd ed. New York: Springer, 360–368.

Manley, D., 2014. Scale, aggregation, and the modifiable areal unit problem. *In*: M.M. Fischer and P. Nijkamp, eds. *Handbook of regional science*. Berlin: Springer-Verlag, 1157–1171.

Manley, D., Flowerdew, R., and Steel, D., 2006. Scales, levels and processes: studying spatial patterns of British census variables. *Computers, Environment and Urban Systems*, 30, 143–160. doi:10.1016/j.compenvurbsys.2005.08.005

Openshaw, S., 1984. The modifiable areal unit problem. Concepts and techniques in modern geography. Norwich: Geo Books, 38.

Openshaw, S. and Taylor, P.J., 1979. A million or so correlation coefficients: three experiments on the modifiable areal unit problem. *In*: N. Wrigley, ed. *Statistical applications in the spatial science*. London: Pion, 127–144.

Qi, Y. and Wu, J., 1996. Effects of changing spatial resolution on the results of landscape pattern analysis using spatial autocorrelation indices. *Landscape Ecology*, 11, 39–49. doi:10.1007/BF02087112

Tiefelsdorf, M. and Griffith, D.A., 2007. Semiparametric filtering of spatial autocorrelation: the eigenvector approach. *Environment and Planning A*, 39, 1193–1221. doi:10.1068/a37378

Wong, D.W.S., 1996. Aggregation effects in geo-referenced data. *In*: S.L. Arlinghaus, *et al.*, ed. *Practical handbook of spatial statistics*. Boca Raton: CRC Press, 83–106.

Wong, D.W.S., 2009a. The modifiable area unit problem (MAUP). *In*: A.S. Fotheringham and P.A. Rogerson, eds. *The SAGE handbook of spatial analysis*. Thousand Oaks: SAGE Publications, 105–123.

Wong, D.W.S. 2009b. Modifiable areal unit problem. In: R. Kitchin and N. Thrift, eds.. *International encyclopedia of human geography*. Amsterdam: Elsevier, Vol. 7, 169–174.

Wong, D.W.S., Lasus, H., and Falk, R.F., 1999. Exploring the variability of segregation index D with scale and zonal systems: an analysis of thirty US cities. *Environment and Planning A*, 31, 507–522. doi:10.1068/a310507

Yoo, E.H., Kyriakidis, P.C., and Tobler, W., 2010. Reconstructing population density surfaces from areal data: a comparison of Tobler's pycnophylactic interpolation method and area-to-point kriging. *Geographical Analysis*, 42 (1), 78–98. doi:10.1111/j.1538-4632.2009.00783.x.

Appendix. A brief note on the results from the hexagon tessellations

Here, the outcome of the same experimental simulation done for the regular tessellation of 1,024 hexagons is presented. Since the spectrum of eigenvectors derived from the hexagon tessellations are not symmetric, nine Moran eigenvectors are chosen with an equal spacing of about 0.20 across the feasible range; SA1 (EV1, MC = 1.0306), SA2 (EV64, MC = 0.8000), SA3 (EV129, MC = 0.6035), SA4 (EV205, MC = 0.3999), SA5 (EV296, MC = 0.1998), SA6 (EV409, MC = −0.0021), SA7 (EV566, MC = −0.1995), SA8 (EV886, MC = −0.3998), and SA9 (EV1024, MC = −1.5186).

Figures A1 and A2 display the MAUP effects, respectively, for the variances and the standardized MCs. Basically, there are no substantial differences; all the results are comparable; minor differences are attributable solely to the differences in the eigenvector spectrums. Accordingly, almost the same set of regression equations is obtained, which can be seen with Table A1.

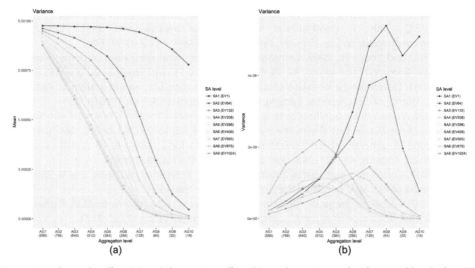

Figure A1. The scale effect (a) and the zoning effect (b) on the variances by the initial level of spatial autocorrelation for the hexagon tessellation.

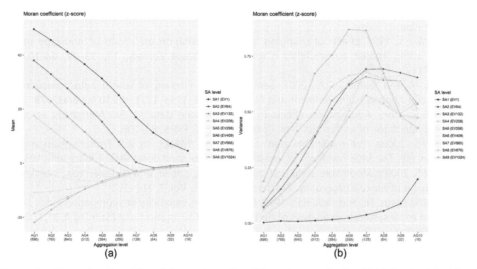

Figure A2. The scale effect (a) and the zoning effect (b) on the standardized MCs by the initial level of spatial autocorrelation for the hexagon tessellation.

Table A1. Regression results for the hexagon tessellations.

	Means		Variances		Observed MCs		Standardized MCs	
	Scale effect (Mean)	Zoning effect (Variance)	Scale effect (Mean)	Zoning effect (Variance)	Scale effect (Mean)	Zoning effect (Variance)	Scale effect (Mean)	Zoning effect (Variance)
Intercept	-2.2349 (-0.136)	10.1751 (0.132)	-0.4734 (-0.110)	1.1877 (0.425)	-1.2788 (-2.731**)	6.0866 (6.183***)	1.2201 (0.757)	0.5876 (11.698***)
AG	-0.0002 (-0.008)	0.0044 (0.037)	0.1056 (15.807***)	0.0090 (2.072*)	-0.0009 (-1.265)	-0.0087 (-5.704***)	-0.0138 (-5.557***)	-0.0006 (-7.449***)
SA	-5.4452 (-0.138)	8.3856 (0.045)	13.1284 (1.266)	-6.1708 (-0.920)	1.8528 (1.650)	0.8781 (0.372)	8.4137 (2.178*)	-0.2129 (-1.767)
D_SA	-7.9224 (-0.325)	-82.7724 (-0.720)	-7.7211 (-1.201)	2.1648 (0.521)	-5.7952 (-8.325***)	0.4760 (0.325)	-22.9336 (-9.575***)	0.2207 (2.955**)
AG*D_SA	0.02819 (0.829)	-0.3465 (-2.163*)	-0.0162 (-1.807)	-0.0281 (-4.838***)	0.0080 (8.027***)	0.0006 (0.294)	0.0463 (13.875***)	0.0001 (0.836)
SA*D_SA	-13.4244 (-0.292)	530.6288 (2.452*)	44.1832 (3.649***)	32.7200 (4.177***)	9.6383 (7.349***)	-2.7713 (-1.005)	24.3805 (5.403***)	-0.1864 (-1.324)
R^2	0.0359	0.3190	0.8841	0.5022	0.8844	0.4538	0.8937	0.6270

Note: 1. Numbers in parentheses are t-values for the regression coefficients. Significance codes: ***: 0.001; **: 0.01; *:0.05.
2. The dependent variables are multiplied by a constant from the smallest 10 to the largest 10^{10} since their data scale is much smaller than the independent variable. Hence the coefficients for a given variable do not reflect the magnitude of differences across the different summary statistics (i.e. mean, variance, observed MC, and standardized MC).

Spatial autocorrelation and data uncertainty in the American Community Survey: a critique

Paul H. Jung ⑩, Jean-Claude Thill and Michele Issel

ABSTRACT

We argue that the use of American Community Survey (ACS) data in spatial autocorrelation statistics without considering error margins is critically problematic. Public health and geographical research has been slow to recognize high data uncertainty of ACS estimates, even though ACS data are widely accepted data sources in neighborhood health studies and health policies. Detecting spatial autocorrelation patterns of health indicators on ACS data can be distorted to the point that scholars may have difficulty in perceiving the true pattern. We examine the statistical properties of spatial autocorrelation statistics of areal incidence rates based on ACS data. In a case study of teen birth rates in Mecklenburg County, North Carolina, in 2010, Global and Local Moran's I statistics estimated on 5-year ACS estimates (2006–2010) are compared to ground truth rate estimates on actual counts of births certificate records and decennial-census data (2010). Detected spatial autocorrelation patterns are found to be significantly different between the two data sources so that actual spatial structures are misrepresented. We warn of the possibility of misjudgment of the reality and of policy failure and argue for new spatially explicit methods that mitigate the biasedness of statistical estimations imposed by the uncertainty of ACS data.

Introduction

The American Community Survey (ACS) data were first released by the U.S. Census Bureau in 2006. Since then, the ACS data has increasingly been used for place-based research across various disciplines, including geography and public health. Given that the ACS data are collected annually, they are particularly useful in generating demographic profiles of fine-grained neighborhoods, such as census tracts, and to track changes in those socioeconomic profiles over time. The use of ACS data for planning and decision making has accelerated with the growing availability of spatial analysis and geographic information (GIS) tools. The resulting body of place-based research using small-area demographic information can now investigate such fine-scale spatial dynamics as neighborhood change, health-related incidences and crimes based on ACS population estimates.

The fine temporal granularity of ACS offers clear advantages over decennial census data, but a distinctive trade-off occurs when using small area estimates of ACS. Due to small sample sizes at the census tract level, ACS variables tend to have large error margins. The high margins of error both complicate data analysis and decrease confidence in the estimates. The National Research Council (2007) recommended that 12 percent or less of the coefficient of variation is an acceptable level of sampling error, but it has been found that most estimates at census tract and block group levels fail to pass this criterion (Spielman *et al.* 2014). Recent evidence (Spielman *et al.* 2014, Folch *et al.* 2016) further indicates that spatial and structural variations of data uncertainty exist across surveyed neighborhoods. Thus, the small area observations derived from the ACS data may not accurately depict the demographic profiles of neighborhoods, leading to incorrect assessments and inappropriate subsequent decisions.

Nonetheless, ACS data are widely used in geographic, public health, and socioeconomic policy studies. In particular, Moran's I and local indicators of spatial association (LISA) are instrumental in detecting spatial dependence in health-related events or disease incidence and prevalence. Demographic profiles like total population counts, race, gender and birth counts, are often used as denominator data to calculate the area prevalence rate in these analyses. Although the U.S. Census Bureau provides detailed information on ACS data error margins for each estimate and guidelines for measuring data quality, we question the extent to which the low data quality of small area estimates in ACS is taken into consideration in the real-world policymaking and academic research, and whether such analyses have the requisite statistical validity to detect the accurate social profiles when relying on ACS data. Statistics can be severely distorted, especially at the resolution of census tracts, where the estimates have the highest sampling errors. Spatial scientists have worked to unravel the uncertainty issue of ACS, including statistical properties stemming from its survey structure and the magnitude and heterogeneity of data uncertainty across areas (Spielman and Folch 2015, Spielman and Singleton 2015, Folch *et al.* 2016). However, much less attention has been devoted to developing approaches to attenuate high uncertainty of ACS data through statistical estimation, except a few methodological developments by statisticians (Porter *et al.* 2015, Bradley *et al.* 2016).

In this paper, we take the stance that the use of ACS data in spatial autocorrelation statistics is highly problematic without considering error margins and we argue for comprehensive awareness of various aspects of ACS data uncertainty and of the effect of this uncertainty in real-world problems and interventions. To this end, we synthesize different aspects of ACS data uncertainty and its implications as they pertain to spatial analysis. We also critically evaluate the structure of ACS and statistically examine how spatial autocorrelation statistics are statistically biased when variables have high sampling errors as ACS data are used.

The rest of this paper is organized as follows. First, ACS is described in some detail with a focus on its survey methodology and discussion of error margins and uncertainty issues. Then, we argue that uncertainty in ACS data have the potential to invalidate statistical inference of spatial autocorrelation in community- and neighborhood-based empirical studies of public health. We analytically study the attenuation of spatial autocorrelation estimates by sampling errors in variables. Finally, we present a case study of teen birth rates calculated on ACS data in Mecklenburg county, North Carolina, to illustrate how ACS biases spatial autocorrelation estimates. The results of this analysis are evaluated in comparison to the ground truth values from decennial census data. The last section concludes and sets the agenda for future research.

Spatial data uncertainty of American Community Survey

Overview of American Community Survey

In 2010 the ACS completely replaced the decennial census long-form survey (Macdonald 2006). The long-form census survey that preceded ACS asked sampled households detailed information like marital status, educational attainment, ancestry, income, work status, number of rooms and vehicle availability (U.S. Census Bureau 2000). In the 2000 U.S. Census, one in six households participated in this survey (U.S. Census Bureau 2002). Each long-form survey data record is used to produce estimates for the surveyed year, so there is a one-to-one match between survey observation and estimates. However, these data have critical shortcomings, notably needing to wait 10 years to have updated demographic information in small areas.

To address this issue, a rolling sampling design started to be applied to ACS instead. This new technique is based on Kish's work on continuous measurement (Alexander 2002). The basic idea behind this design is that repeated survey on non-overlapping groups with smaller sample sizes can generate estimates over multiple years (Kish 1990). Rather than increasing the sample size in a single period, it divides the sample sizes over months or years and increases the number of surveys. Accumulation of monthly survey data reduces cost and enables the calculation of estimates on a yearly basis. With a temporal moving window, multiyear pooled survey data are used for annual estimates. The window sizes vary with the granularity of geographical units. In case of small areas like census tracts where having a sufficient sample sizes can be a challenge, annual estimates derive from 5-year pooled survey data. If the geographical unit has a population over 65,000, single-year survey data are used for estimates in each year (U.S. Census Bureau 2008). By incorporating a rolling sampling design and using multiyear pooled data, ACS can overcome the critical weakness of decennial census data.

Sampling error and structural uncertainty

The rolling sampling technique permits the provision of annual demographic data at a fine spatial granularity. ACS data can now be 'warmer' (more current) than the decennial census, but also 'fuzzier' (less precise) (Macdonald 2006). Thus, ACS data are more uncertain than decennial census long-form estimates, even though they provide frequent estimates in intercensal years.

The uncertainty of ACS estimates stems from sampling errors and non-sampling errors. On the latter, ACS has better accuracy than the census long-form survey because ACS employs a smaller number of well-trained professional staffs who can more effectively trace non-respondents (U.S. Census Bureau 2004, Macdonald 2006). Thus, the higher uncertainty of ACS estimates is squarely attributable to higher sampling errors introduced by rolling sampling designs and limited sample sizes.

The high uncertainty is mainly because ACS uses much smaller national samples than the decennial survey. Whereas the 2000 decennial census long-form surveys one of six households (17%) in one year, ACS only takes 12.5% of all addresses over 5 years. The U. S. Census Bureau expected that ACS would exhibit 33% more uncertainty than the decennial census survey, but earlier study suggests that the uncertainty would be 50% greater based on sample size assumptions (Starsinic 2005). ACS 5-year estimates of 2005–2009 are found to have about 75% greater uncertainty than the decennial census (Navarro 2012). Furthermore, the response rate of ACS tends to be lower than that of the decennial census long-form survey (Hough and Swanson 2004, Salvo et al. 2004, Gage 2006), which is attributable to the higher uncertainty (Macdonald 2006). The uncertainty problem is especially severe when small area estimates are concerned, owing to the lower sample sizes at local geographies on average compared to the decennial census survey (Spielman et al. 2014, Folch et al. 2016). Despite efforts by U.S. Census Bureau to reduce the high variances of ACS estimates in ACS through statistical strategies like the GREG procedure (Fay 2007), such treatment still does not cure their high uncertainty enough, as U.S. Census Bureau found that the 5-year estimates can have even larger variances than the corresponding 1-year estimates (Griffin and Albright 2015).

Besides the small sample sizes that characterizes ACS at the tract level, the high uncertainty by sampling error is also attributed to survey tract design and resulting population heterogeneity within survey tracts (Spielman et al. 2014). When the sampled population has more diverse demographic characteristics, such as race, income, educational attainment, it is more likely that the group statistics deviate from corresponding characteristics in the population group and each sampling produces inconsistent estimates (Rao and Molina 2015). This issue is exacerbated with small sample sizes.

These issues scale up to a structural variation of uncertainty across census tracts associated with demographic structures of cities. With regard to median income measures, the reliability of ACS estimates is found to vary according to the distance from the urban center and to the poverty level of census tracts (Spielman et al. 2014). There are robust tendencies in U.S. metropolitan areas for inner city areas and poor neighborhoods to have higher uncertainties than outer areas and affluent neighborhoods. Also, census tracts with more non-white residents and higher racial segregation tend to have greater uncertainty (Folch et al. 2016). On-site case study in Chestnut Hill, Tennessee also confirms that ACS does not represent marginalized neighborhoods well and that policy making based on ACS data become problematic in such areas (Bazuin and Fraser 2013). Considering that inner-city areas and non-White neighborhoods tend to have heterogeneous population, these structural patterns in data uncertainty are strongly related to how the tract populations are surveyed for the ACS. This issue has been acknowledged in a similar context by Logan et al. (2018) where it is argued that segregation measures using ACS estimates may lead to wrong conclusions, so that there is compelling need for correction by their uncertainty levels.

These findings present important implications to place-based health research. Considering that health-related phenomena are highly associated with socioeconomic status and races, the variation of uncertainty of ACS data may cause ACS-driven research to provide a biased view while the true problem in marginalized neighborhoods may not be well acknowledged. Therefore, public health studies should be extremely cautious in using small-area estimates of ACS and should recognize and mitigate uncertainty's higher levels and deeper spatial variability.

The U.S. Census Bureau (2008) officially stated the importance of considering the uncertainty of ACS estimates and its possible impacts on statistical significance. For this reason, the values of error margins are provided along with the estimates in each reporting area. The reported error margins pertain to a 90% confidence interval, which is 1.645 times of the standard error, from which sampling error can be readily measured by the coefficient of variation (CV). This information can be used for inferential purposes to determine if the statistics are reliable enough. Even though ACS does not provide unbiased and consistent estimates, error margins enable users to determine how reliable reported estimates are when examining socioeconomic change of neighborhoods or designing a place-based policy.

It is critically important for spatial researchers to be aware of the scope of discrepancies between small-area estimates reported by ACS and the true demographic features, and that ACS data may not always be an accurate source of demographic information. Nevertheless, there is only scant public health study that carefully considers issues of high uncertainty in ACS. A scan of applied and policy research raises concerns due to their lack of awareness of statistical uncertainty, of its ensuing propagation, and of the distortion of the understanding of demographic landscapes, in spite of the effort of the U.S. Census and of the community of spatial scientists to publicize the data uncertainty issue in ACS (Jurjevich et al. 2018). Considering that many health-related phenomena are associated with poverty and race composition in neighborhoods, data uncertainty may distort our understanding of their true geospatial aspects and cause biases in place-based policies.

The attenuation of spatial autocorrelation estimates by sampling errors in variables

We examine here the statistical properties of spatial autocorrelation statistics of areal incidence rates under high and heterogeneous data uncertainty. Our focus is particularly on how much the statistical estimation of Global and Local Moran's I is diluted when each area has heterogeneous sampling errors. Let us consider N areas indexed by i that constitute a study region. The incidences are measured by area, so that each area i has an incidence rate θ_i, and a raw rate $x_i = r_i/p_i = \hat{\theta}_i$, which is a maximum likelihood estimator of θ_i where r_i is a count of cases and p_i is the population-at-risk. The true Global Moran's I is as follows:

$$I = \frac{N}{\sum_{ij} w_{ij}} \frac{Z'WZ}{Z'Z} \tag{1}$$

where Z is a $N \times 1$ vector whose element z_i is the deviation from the global average rate m, $z_i = \theta_i - m$, $m = \sum_i r_i / \sum_i p_i$ and W is a $N \times N$ spatial weight matrix that defines the spatial relationship between i and j by w_{ij}. When W is row-standardized, then (1) boils down to $I = Z'WZ/Z'Z$.

Here we consider two cases: (1) where the population-at-risk p_i cannot be observed and the estimate n_i observed by sampling survey is used for computing the raw rate x_i;

(2) neither incidence count r_i nor population-at-risk p_i are observed and the areal incidence rate estimate x_i is observed through sampling survey instead for θ_i. Hence, in the former case, the observed incidence rate in area i is $x_i = r_i/n_i$, where r_i is taken from a separate local enumeration survey without sampling error, but we take estimate x_i for θ_i in the latter case assuming both r_i and n_i are unknown. Thus, either p_i or x_i is taken from the sampling-survey estimates, such as ACS. We assume that $x_i = \theta_i + \varepsilon_i$, where ε_i is independently distributed, $\varepsilon_i \sim (0, \sigma_i^2)$ and uncorrelated with others and with the estimates θ_i, $E(\theta_i \varepsilon_i) = 0$ for all i. That is, rate x_i is distributed around the true incidence rate θ_i but the accuracy varies across areas without correlation to the true incidence rate θ_i.

Since p_i and θ_i cannot be observed directly, the true deviation z_i and Global Moran's I cannot be observed either. Let us assume that Z is approximated by an observed deviation vector Z^* in which each element $\hat{z}_i = x_i - \hat{m}$ and $\hat{m} = \sum_i r_i / \sum_i n_i$. m can be approximated by $\hat{m}$ when sufficient areas are aggregated, so that $\hat{m} = m$. Hence, $Z^* = Z + \varepsilon$. The observed Global Moran's I is as follows:

$$I^* = \frac{Z^{*\prime} W Z^*}{Z^{*\prime} Z^*} = \frac{(Z + \varepsilon)^\prime W (Z + \varepsilon)}{(Z + \varepsilon)^\prime (Z + \varepsilon)} = \frac{(Z^\prime W Z + Z^\prime W \varepsilon + \varepsilon^\prime W Z + \varepsilon^\prime W \varepsilon)/N}{(Z^\prime Z + Z^\prime \varepsilon + \varepsilon^\prime Z + \varepsilon^\prime \varepsilon)/N} \tag{2}$$

The independence assumptions on the measurement error imply that $E(\varepsilon_i \varepsilon_j) = 0$, $E(\varepsilon_i z_i) = 0$ and $E(\varepsilon_i z_j) = 0$ for all i and j, so we take 0 for $Z^\prime W \varepsilon/N$, $\varepsilon^\prime W Z/N$, $\varepsilon^\prime W \varepsilon/N$, $Z^\prime \varepsilon/N$ and $\varepsilon^\prime Z/N$ by the asymptotic equivalence. This gives

$$I^* = \frac{Z^\prime W Z}{Z^\prime Z + \varepsilon^\prime \varepsilon} = \frac{Z^\prime Z}{Z^\prime Z + \varepsilon^\prime \varepsilon} \times \frac{Z^\prime W Z}{Z^\prime Z}. \tag{3}$$

When we apply asymptotic equivalence, $E(Z^\prime Z/N) = Var(\theta)$ and $E(\varepsilon_i^2) = Var(\varepsilon_i) = \sigma_i^2$. This gives

$$I^* = \frac{Var(\theta)}{Var(\theta) + Var(\varepsilon)} \times I = \frac{Var(\theta)}{Var(x)} \times I \tag{4}$$

where $Var(\varepsilon) = \sum Var_i(\varepsilon_i)/N = \sum \sigma_i^2/N$. We should note that $I^* \leq I$, and I^* is downward biased by $Var(\theta)/Var(x)$, which is the separation index or reliability coefficient. This is a special case of regression dilution bias that occurs when the predictor variable has known measurement error (Fuller 1987, Frost and Thompson 2000). This inequality implies that measurement from large sampling errors causes attenuation of Global Moran's I estimates. Especially when variance of the observed area incidence rate is sufficiently greater than the true incidence rate, such as with small-area estimates like ACS, it potentially distorts the observation of the global spatial autocorrelation structure of the study region.

Derived from the Global Moran's I, the Local Moran's I (Anselin 1995) of the true area incidence rate is defined by

$$I^i_{local} = \frac{N}{Z'Z}\left(z_i \sum_j w_{ij}z_j\right) \tag{5}$$

or in matrix form,

$$I^i_{local} = \frac{diag(WZZ')}{Z'Z/N} \tag{6}$$

The true deviation Z is substituted by the observed deviation Z^* when area incidence rates have measurement errors; thus, the observed Local Moran's I is as follows:

$$
\begin{aligned}
I^*_{local} &= \frac{diag\left(WZ^*Z^{*'}\right)}{Z^*Z^*/N} = \frac{diag\left(W(Z+\varepsilon)(Z+\varepsilon)'\right)}{(Z+\varepsilon)'(Z+\varepsilon)/N} \\
&= \frac{N \times diag(W(ZZ' + Z\varepsilon' + \varepsilon Z' + \varepsilon\varepsilon')/N)}{(Z'Z + Z'\varepsilon + \varepsilon'Z + \varepsilon'\varepsilon)/N}
\end{aligned}
\tag{7}
$$

We also apply the asymptotic equivalence of the error-related terms, $E(z_i\varepsilon_i) = 0$, and we take 0 for $Z\varepsilon'$ and $\varepsilon Z'$. Also, the covariance matrix of the error is $E(\varepsilon\varepsilon') = \Omega$ where Ω is a $N \times N$ diagonal matrix in which each ith diagonal element is σ_i^2, so we take Ω for $\varepsilon\varepsilon'$. This boils down to

$$I^*_{local} = \frac{diag(W(ZZ' + \Omega))}{(Z'Z + \varepsilon'\varepsilon)/N} = \frac{Z'Z}{Z'Z + \varepsilon'\varepsilon} \times \frac{diag(WZZ')}{Z'Z/N} + \frac{\varepsilon'\varepsilon}{Z'Z + \varepsilon'\varepsilon} \times \frac{diag(W\Omega)}{\varepsilon'\varepsilon/N} \tag{8}$$

We can apply again the asymptotic equivalence of variances, $E(Z'Z/N) = Var(\theta)$ and $E(\varepsilon_i^2) = Var(\varepsilon_i) = \sigma_i^2$. This gives

$$I^*_{local} = \frac{Var(\theta)}{Var(x)} \times I_{local} + \frac{Var(\varepsilon)}{Var(x)} \times \frac{diag(W\Omega)}{Var(\varepsilon)} \tag{9}$$

Here $diag(W\Omega)_i$ stands for the spatial-weighted sum of the error variances of the areas adjacent to i. It should be noted that the diluted Local Moran's I I^*_{local} is the weighted sum of the I_{local} and $diag_i(W\Omega)/Var(\varepsilon)$, implying that the level of attenuation is affected both by the overall reliability ratio $Var(\theta)/Var(x)$ and by the ratio of spatial-weighted error variance of area i to the overall error variance $diag_i(W\Omega)/Var(\varepsilon)$. The measurement errors are heteroscedastic, so the attenuation of the estimates varies with the level of uncertainty across areas. We can consider two directions how Local Moran's I values can be diluted. First, in case one area's neighboring areas have low measurement errors, $diag_i(W\Omega) \ll Var(\varepsilon)$, the Local Moran's I is deflated from the true value, possibly causing Type I (False Positive) error. Second, one area's neighbors may have high measurement errors $(diag_i(W\Omega) \gg Var(\varepsilon))$, in which case Local Moran's I is inflated to $diag_i(W\Omega)/Var(\varepsilon)$. This may bring Type II (True Negative) error. Either case shows that measurement errors of areal rates that are high and heteroscedastic can cause significant discrepancy between the true local spatial autocorrelation structure and the observed one.

Overall, the statistical examination of Global and Local Moran's I statistics corroborates that data uncertainty influences the statistical estimation and challenges the validity of spatial autocorrelation detection with these statistical tools. The issue of attenuation by error-in-variables has been acknowledged in psychology and health sciences, where the measurement error usually occurs during the psychometric survey and clinical experiments, and methodological adjustments have been proposed within this context (Smith and Phillips 1996, Frost and Thompson 2000, Hutcheon *et al.* 2010, Dima 2018). However, this issue has remained rather elusive in socioeconomic studies using small-area demographic data, including geography and place-based public health. Considering that the data uncertainty issue of ACS now challenges place-based research, we argue it is timely to revisit the statistical treatment of the attenuation by uncertain socioeconomic data like ACS.

Case study: teen birth rates in Mecklenburg county, north Carolina

Data and areal incidence rate estimates on ACS

In this section, we present a case study that exemplifies the problematic use of small-area data with high uncertainty, like ACS estimates, specifically in spatial autocorrelation detection. We examine the spatial variation of census-tract teen birth rates in Mecklenburg County, North Carolina. A data file containing the list of females aged 15 and 19 who had given birth in 2010 was acquired from the Mecklenburg County Department of Public Health. After geocoding the data, the number of births was aggregated by census tract based on the mothers' georeferenced address information. The data aggregated by census tract effectively deidentifies individual information of each birth record. Then, the aggregated incidence count data are linked to demographic data from the ACS and decennial census.

We use 5-year 2006–2010 ACS estimates, which minimize sampling errors among different time windows, and 2010 decennial census statistics which are used to provide ground truth for benchmarking purposes. We take the results on the decennial census statistics as the benchmark since the decennial census is a complete enumeration without sampling errors, while ACS estimates have both sampling errors and non-sampling errors. We can consider the 2010 decennial census statistics are the best benchmark for corresponding ACS estimates despite its non-sampling error, because the high uncertainty in ACS estimates is largely attributable to their large sampling errors (Macdonald 2006).

The teen birth rates are calculated as the ratio of the teen birth incidence counts to female population aged between 15 and 19. We compute three rates for comparison purposes: (1) Teen birth rate calculated on the birth certificate records as numerator and decennial census population statistics as denominator (census raw rate), (2) Fertility rate estimates of women aged between 15 and 19 in ACS (ACS fertility) and (3) teen birth rate calculated on the birth certificate records as numerator and ACS population estimates as denominator (ACS raw rate). Thus, the first case has no sampling error; the second case considers sampling errors of both numerator and denominator, while the latter assumes those only in the denominator.

The three teen birth rates are depicted graphically through choropleth maps with quintile breaks classification of census raw rate for comparison (Figure 1(a-c)). The spatial disparity of teen birth rates is distinctive and the spatial trend is broadly consistent

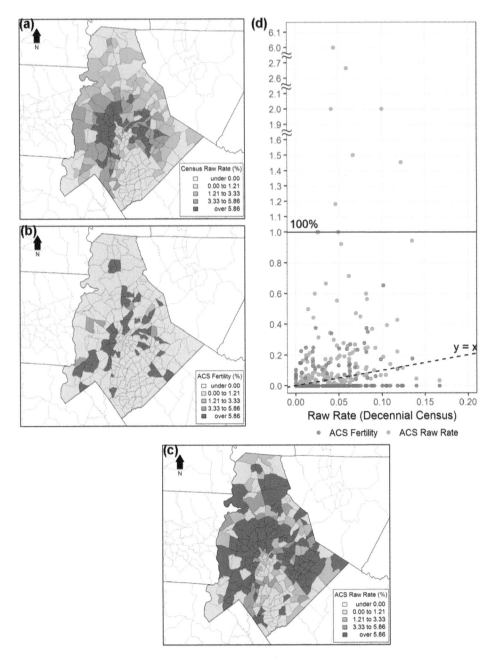

Figure 1. Census-tract level map of teen birth rates and comparison of teen birth rates created on decennial census (2010) and ACS (2005–2010): (a) Benchmark choropleth map of census raw rates of teen birth created on birth certificate records and decennial census; (b) Fertility rate estimates in ACS; and (c) ACS Raw rates created on birth certificate records and ACS; (d) Scatter plot between census raw rates, acs fertility estimates and acs raw Rates. The dashed line is the equality line, where rates are equal.

between the three choropleth maps. In the three maps, we can identify a stark delineation between a southern wedge-shaped area with low teen birth rates and an inner crescent with higher rates around Uptown (central business district) of Charlotte.

Closer examination reveals important differences, however. With the breaks classification set identically across the three maps, the ACS fertility estimates (Figure 1(b)) and ACS raw rates (Figure 1(c)) show stark departure from the benchmark statistics on decennial census (Figure 1(a)). While ACS fertility estimates tend to be underestimated, ACS raw rates are broadly inflated, with an enlarged crescent with higher rates around Uptown of Charlotte. This also reveals that the distortion of teen birth rates by ACS is more severe in census tracts with higher rates. Thus, choropleth mapping of areal incidence rates on ACS may distort the readers' perception of spatial variations and areas with higher rates should be examined with great care. The scatter plot between the two rates confirms different bounds of ranges (Figure 1(d)). While the census raw rate ranges between 0 and 16.7%, the ACS fertility estimate ranges between 0 and 65.3%. The ACS raw rate is widely dispersed between 0 and 600%, including 51 census tracts above the maximum of the decennial census rate, 16.7%. These highly inflated or deflated rates result from the underestimation of the denominator, women population aged between 15 and 19, or the fertility rates of mothers aged between 15 and 19, by ACS.

Since ACS incorporates a rolling sampling survey structure, the large sampling errors at the census-tract level may cause such large magnitudes of errors of areal incidence rates. The plot in Figure 1(d) shows that most ACS fertility estimates are deflated to zero and some are overestimated with respect to the corresponding census raw rates, whereas the ACS raw rate tends to be inflated with respect to the corresponding census raw rate. Differences in the deviation from the equality line confirm that the spatial variation of sampling errors of ACS estimates may cause the heteroscedastic pattern of the errors between the ACS and decennial census rates (Spielman *et al.* 2014, Folch *et al.* 2016).

Log-scaled scatter plots present that the magnitudes of errors are closely associated with the respective margin of errors of ACS estimates, confirming that the high error rate of ACS rates results from data uncertainty in ACS (Figure 2). Since most ACS fertility estimates are zero, we use their standard errors, obtained as margin of errors divided by 1.645, to evaluate their data uncertainty instead of CVs. The errors of ACS fertility estimates are distributed weakly proportionally to the magnitude of the standard errors,

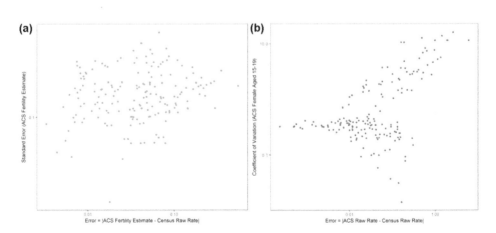

Figure 2. Log-scaled scatter plots between errors of incidence rates and margin of errors of ACS estimates: (a) Error of ACS fertility estimate and standard error; (b) Error of ACS raw rate and coefficient of variation of female population aged between 15 and 19 in ACS.

confirming that tracts with high errors tend to have higher standard errors of ACS fertility estimates (Figure 2(a)). The proportional pattern between the observed errors of ACS raw rate and their CVs also shows that the large errors result from high uncertainty in denominator data (Figure 2(b)). Despite some irregular patterns found in the plots, large observed errors on the two rate estimates seem associated with high standard errors of ACS fertility estimates or CVs of denominator data.

Finally, the overall accuracy of the ACS rates can be evaluated for census tracts by the separation index or reliability coefficient (Fuller 1987), $Var(\theta)/Var(x)$ where x is a vector of either of the observed two rates on ACS and θ is the benchmark true rate on decennial census. The separation index of ACS fertility estimates is 15.23% and that of ACS raw rates is only 0.43%. This strongly suggests that the incidence rates created on ACS are too unreliable to represent the true areal incidence of teen birth in census tracts.

Global spatial autocorrelation detection under data uncertainty

In order to assess the statistical validity of spatial autocorrelation statistics of the incidence rates with high and heterogeneous uncertainty, we compare the Moran's I spatial autocorrelation statistics on the three rates based on ACS and decennial census. Since the exact distributional properties of Moran's Is cannot be established, we follow the practice set by Cliff and Ord (1981) and Anselin (1995) and use a Monte Carlo permutation test to evaluate statistical significance. We perform 10,000 random permutations on Moran's I and produce pseudo p-values. The spatial weight matrix is specified by Queen's contiguity.

We first examine the independence and uncorrelatedness among measurement errors and incidence rates to check if the data in our case study can be applied to the mathematical derivation in the previous section. Thus, we check that the asymptotic equivalences hold for $E\left(\varepsilon_i \sum_j W_{ij}\varepsilon_j\right) = 0$, $E(z_i\varepsilon_i) = 0$ and $E\left(z_i \sum_j W_{ij}\varepsilon_j\right) = 0$ for all i and j. Scatter plots between errors of teen birth rates (ε) and either census raw rate (z), spatial lag of census raw rate $\left(\sum_j W_{ij}z_j\right)$ or spatial lag of the errors $\left(\sum_j W_{ij}\varepsilon_j\right)$ do not present any linear or structural relationships among them but their covariances are approximated to zero (Figure A1 in Appendix). This confirms that $E\left(\varepsilon_i \sum_j W_{ij}\varepsilon_j\right) = 0$, $E(z_i\varepsilon_i) = 0$ and $E\left(z_i \sum_j W_{ij}\varepsilon_j\right) = 0$ hold for all i and j. This implies that Global and Local Moran's Is are significantly attenuated and biased by errors-in-variables as derived in Eq. (4) and (9).

Global Moran's I of ACS fertility estimate and ACS raw rate is 0.128 and 0.025, respectively, which can be compared to that of the decennial census rate, 0.46 (Figure 3(d)). This verifies that Global Moran's I on sampled estimates with high uncertainty, such as census-tract level ACS estimates, is biased downward by the reliability coefficient, $Var(\theta)/Var(x)$ in Equation (4). Since the reliability coefficient of ACS fertility estimate is 15.23%, its Global Moran's I, 0.128, is in fact close to $0.46 \times 15.23\% = 0.07$. Global Moran's I of ACS raw rate, 0.025, is also approximated to $0.46 \times 0.43\% = 0.002$. Whereas the pseudo p-value of Moran's I of ACS fertility estimate and ACS raw rate is 0.19% and 14.91%, respectively, the decennial census rate is less than 0.00%, which indicates a stark discrepancy in statistical significance between the results from ACS estimates and benchmark result from decennial census. Moran plots also show how Moran's Is on both ACS rate estimates are biased and lose statistical significance. A small number of outlier tracts with high errors have heavy influence on the spatial autoregressive line and sharply tilt the spatial autocorrelation trend (Figure 3(b-c)). In contrast, the spatial autocorrelation trend of the decennial census rate is robust across the range of the rate, with no outlier to leverage the spatial autoregressive line (Figure 3(a)). These results corroborate that high sampling errors of denominator data result in a regression dilution bias of Moran's I, which leads to a high risk of Type I error.

The performance of the spatial autocorrelation statistic under high and heterogeneous data uncertainty can be evaluated through a formal test. Our analytical strategy is to compare Moran's I statistics of ACS fertility estimate and ACS raw rate in 2010 to the ground truth benchmark. The discrepancies of the former two statistics with Moran's I of census raw rate measure the risk of misjudgment on the detected spatial autocorrelation structure. We perform an inequality test between the Moran's I values of either ACS fertility estimate or ACS raw rate (I_{ACS}) and of census raw rate (I_{DC}) by means of the following two-tailed hypothesis test

$$H_0 : I_{ACS} = I_{DC}$$

$$H_1 : I_{ACS} \neq I_{DC}.$$

The difference between the two Moran's Is can be tested based on the assumption of normal distribution of the two statistics, but there is no ground for the a priori assumption of normality, as noted earlier. Following the same approach used to estimate Moran's I, we perform a two-tailed Monte Carlo permutation test to evaluate the statistical significance of the equality of the two Moran's I statistics. The test statistics is $\hat{I}_{ACS} - \hat{I}_{DC}$. The pseudo p-value is computed from 10,000 permutations.

The inequality tests (Figure 3(e-f)) show that the Moran's I statistics on the ACS rates are statistically underestimated and biased downward with respect to the benchmark value on the decennial census rate. The test statistic completely deviates from the null distribution of random permutations with pseudo p-value of 0. Thus, since the global-scale spatial autocorrelation structure detected in both ACS fertility estimates and ACS raw rates cannot approximate that of the census raw rate, it is not possible to capture the true spatial autocorrelation structure when using ACS data. This provides statistical evidence of regression dilution and downward biasedness of Global Moran's I by sampling errors in ACS data, as provided in Equation (4) and Figure 3(a-c).

Local spatial autocorrelation under data uncertainty

The comparison of LISA maps based on the ACS and decennial-census rates, respectively, is conducted following the same principles as for the Global Moran's I. Consistently with the global scale analysis, results at the local scale point that using ACS critically diminishes statistical validity. The benchmark LISA map of the decennial-census rate finds several clusters, specifically hot spots around the inner crescent with higher rates and cold spots

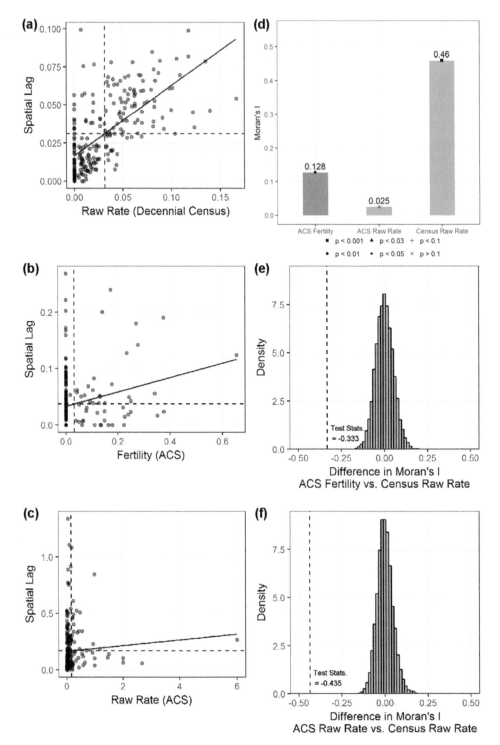

Figure 3. Moran plots and teen birth rates and inequality test of global Moran's I: (a) Moran plot of census raw rate; (b) ACS Fertility estimates; (c) ACS raw rate; (d) Moran's I of three rates; (e) random permutation test on difference in global Moran's I between ACS fertility estimates and census raw rate; (f) Difference in ACS raw rate and census raw rate (Note: Global Moran's I and difference in Global Moran's I are generated by 10,000 random permutations).

in outer southern areas as well as in one tract in the north at 5% significance level (Figure 4(a)). This result is consistent with the color-graded spatial pattern found in choropleth maps of rates (Figure 1(a-b)). However, Local Moran's Is of ACS fertility estimate detect only a few hot spots at the 5% significant level, which are located in inner southwestern areas (Figure 4(b)). This would lend support to the existence of a weak spatial autocorrelation structure unlike that of census raw rate. In a similar way, Local Moran's Is of ACS raw rate fails to identify any hot or cold spots (Figure 4(c)) in tracts where the benchmark LISA map detected such clusters, and even detects two hot spots in tracts classified as not-significant in the benchmark LISA map. Clearly, the LISA maps of both ACS fertility estimate and ACS raw rate fail to capture the local-level spatial autocorrelation trend of the benchmark map. Thus, users may misjudge the spatial autocorrelation trend not only at global scale, but also at the local scale.

We follow the same strategy to statistically compare the spatial series of Local Moran's Is created on the decennial census and ACS as for the Global Moran's I. In detail, the inequality tests on each census tract are one-tailed hypothesis tests

$$H_0^i : I_{ACS}^i = I_{DC}^i$$

$$H_1^i : I_{ACS}^i > I_{DC}^i$$

$$H_2^i : I_{ACS}^i < I_{DC}^i$$

where I_{ACS}^i and I_{DC}^i are Local Moran's I of tract i of either ACS fertility estimate and ACS raw rate and of census raw rate, respectively. H_1^i is the alternative hypothesis that I_{ACS}^i is inflated by the high sampling error of ACS in the corresponding tract, while H_2^i indicates that I_{ACS}^i is deflated. Likewise, the test statistics is $\hat{I}_{ACS}^i - \hat{I}_{DC}^i$, and $\hat{I}_{ACS}^i$ and $\hat{I}_{DC}^i$ of each tract are computed based on each permutation. H_1^i and H_2^i are accepted when $\hat{I}_{ACS}^i - \hat{I}_{DC}^i > 0$ and $\hat{I}_{ACS}^i - \hat{I}_{DC}^i < 0$ are statistically significant, respectively. Statistical inference is based on 10,000 permutations.

The inequality tests of Local Moran's Is overall provide the same results as for global-scale measures, namely that sampling errors in either fertility rate estimates or denominator reduce the statistical validity to capture local-level spatial autocorrelation structure. We find that Local Moran's Is are underestimated (H_1) in most census tracts that are detected as hot or cold spots (Figure 4(d-e)). Therefore, the discrepancies of local spatial autocorrelation structures of the two LISA maps of ACS rates with the benchmark from the census raw rate (Figure 4(b-c)) may be imputed to the downward biasedness of Local Moran's I statistics. Since hot and cold spots have relatively higher value of Local Moran's I than other tracts, the deflation of Local Moran's Is has them identified as having no significant local spatial autocorrelation. This result critically confirms Equation (9) that the biasedness of Local Moran's I results from the overall accuracy of the rate, the reliability coefficient ($Var(\theta)/Var(x)$) and high sampling errors in neighboring tracts. Considering that most identified tracts are underestimated, the biasedness of Local Moran's Is is a strong predictor of how low the reliability coefficient due to sampling errors may be. In summary, the inequality tests of Local Moran's I prove that examining Local Moran's I with ACS data can limit

Figure 4. LISA maps of teen birth rates and inequality test of local Moran's I: (a) LISA map of census raw rate; (b) ACS fertility estimates; (c) ACS raw rate; (d) Map of difference in LISA between ACS fertility estimates and census raw rate; (e) Difference between ACS raw rate and census raw rate (Note: Local Moran's is and difference in local Moran's is are tested at 5% significance by 10,000 random permutations).

the perception of the true spatial autocorrelation structure and that high sampling errors of ACS should be taken into account as a factor that diminishes the statistical validity of the Local Moran's I as well as of its global counterpart.

Discussion on existing statistical remedies for ACS uncertainty

Data uncertainty in ACS gives rise to the issue of attenuation or propagation of errors in statistics. While geographers have sought better ways to communicate uncertain demographic information, such as developing geovisualization (Sun and Wong 2010, Sun *et al.* 2015, Wei and Grubesic 2017, Wei *et al.* 2017) and regionalization methodologies (Spielman and Folch 2015, Sun and Wong 2017), statisticians have focused on the effect of errors in variables on the validity of statistical estimation and on developing estimation methodologies to adjust biasedness and improve estimation efficiency. In a broad statistical sense, the interest in error attenuation stems from determining how the magnitude and direction of model estimates are diluted by sampling and measurement errors and how statistical estimation can be corrected to approximate the true parameters. Econometricians also have acknowledged the issue of attenuation by errors in variables and have developed various estimation strategies, such as instrumental variable treatments, general method of moments (GMM) estimation and nonparametric methods (Hausman 2001, Erickson and Whited 2002, Schennach 2004). While the issue of errors-in-variables has received much attention across various fields, spatial scientists have largely shied away from statistical approaches to handle uncertainty in ACS data.

Among the statistical treatments developed to handle the issue of errors-in-variables, methodologies for small area estimation have many features that relate to data uncertainty in small areas like ACS. Small-area estimation methods (Ghosh and Rao 1994, Rao 2005, Datta *et al.* 2011, Rao and Molina 2015) may improve estimates in statistical models when the focus is on the overall trend of areal incidence rather than individual tracts. These methodologies mainly involve Bayesian approaches, which estimate an a priori distribution from the observed data, such as empirical Bayes (EB), hierarchical Bayes (HB) or empirical best linear unbiased prediction (EBLUP) (Ghosh and Rao 1994). They basically use benchmark information from the larger geographic area to adjust estimates in small areas (Datta *et al.* 2011). One approach applied to spatial data is that of EB estimators studied by Marshall (1991), Cressie (1992), Kafadar (1996) and Assunção *et al.* (2005). This approach has been found to be particularly well suited for mapping disease and health-related incidence rates. The EB estimators are used to adjust the extreme rates that often exist in small population areas to present a more stationary spatial pattern on the map. This mapping methodology is implemented in GeoDa (Anselin *et al.* 2006) and the EB approach has also been applied to adjust Moran's I for population size heterogeneity (Assunção and Reis 1999). However, EB methodologies are only starting to be available to handle errors-in-variables resulting from high and heterogeneous sampling errors (e.g. Jung *et al.* 2018), which is widespread in ACS.

The EB statistical approach has several advantages for the treatment of ACS data uncertainty. First, it has solid foundations in statistical theories, both with regard to error attenuation and to small area estimation. Especially, current developments in the EB approach indicate that EB outperforms other approaches (Ghosh and Rao 1994), so it can be seen as a powerful solution for data uncertainty in ACS. Second, the EB statistical approach does not compromise high-resolution information for accuracy, which does occur with spatial aggregation techniques. For ACS users more concerned with the overall spatial trend than precise values of individual areas, the EB adjustment based on uncertainty information in disaggregated geographies is useful as it sharpens the overall spatial trend. Third, developing an estimation methodology on data uncertainty may be the most inexpensive and efficient way to secure both detailed demographic information and high spatiotemporal granularity (Spielman *et al.* 2014). The statistical approach of small area estimation and errors-in-variable models do not require more samples to improve the accuracy of the observations, but it uses existing information on estimates and error margins, as provided by the Census Bureau. Further development in the statistical methodology for ACS data uncertainty would enhance the value of ACS for practical use in place-based empirical studies.

Conclusions

Without question, the ACS has provided a wealth of demographic information in small areas and has invigorated neighborhood-focused studies in many disciplines, such as geography, public health and criminology. However, many researchers have missed or neglected how sampling errors in ACS estimates propagate errors in statistical models. In this paper, we underscored the imperatives of proper statistical treatment to reduce the attenuation of statistical estimation when using data constructed on ACS data, for which a robust understanding of the survey structure and statistical properties of the sampling errors of ACS estimates is required.

ACS estimates have been used in various statistical models at the level of census block groups and tracts, such as spatial and standard regression models, spatial auto-correlation statistics, and in constructing neighborhood indices by combining ACS estimates. However, sampling errors in variables can weaken the validity of statistical models and their conclusions may lose their grounding in empirical evidence, which is seldom acknowledged. Existing geospatial methodologies, such as visualization of the uncertainty information and spatial aggregation techniques, can contribute to alleviat-ing the downsides of the uncertainty of ACS to a certain extent. However, we have argued that they only provide a limited solution since they fail to improve the reliability of statistical estimation and the spatial information from fine spatial granularities is necessarily compromised.

Following a statistical perspective that conceives the sampling errors of ACS as error-in-variables, we shed a new light on small-area sampling errors by examining how sampling errors of ACS estimates attenuate the statistical estimation of Global and Local Moran's I statistics. We warn of the biasedness of the spatial autocorrelation statistics due to sampling errors of ACS estimates and Type II error, and of the failure to detect the true spatial autocorrelation structure of the data. The examination of spatial autocorrelation statistics implies that analytical outcomes based on ACS esti-mates cannot guarantee their statistical validity and reliability. While the analysis per-tained only to spatial autocorrelation statistics, the attenuation of the statistical estimation by sampling errors that was documented for spatial autocorrelation statistics may also critically generate high biasedness in other statistical models like regression models (Folch *et al.* 2016).

Our empirical results expose that the naïve application of ACS data without consid-eration of data uncertainty critically misleads the understanding of the demographic landscape, and beyond this, it may also lead to the failure of local policies based on ACS estimates, especially in block groups or tracts. We argue that neighborhood-based researchers should acknowledge that the use of ACS involves high risk of bias and failure, and we call for a new generation of statistical methodologies that can effectively mitigate the attenuation and biasedness of statistical estimations imposed by the uncertainty in ACS for small areas.

Acknowledgments

This research received approval from the University of North Carolina at Charlotte Institutional Research Board. We wish to acknowledge the work of guest editors for this special issue, Yongwan Chun, Daniel Griffith and Mei-Po Kwan. We appreciate comments made by the anonymous reviewers on earlier versions of the manuscript. They have contributed to enhancing the work reported here in meaningful ways. We also deeply thank Daniel Yonto for help in data acquisition. This work used R 3.3.3 and R Package *spdep, ggplot2* and *GISTools*.

Disclosure statement

No potential conflict of interest was reported by the authors.

ORCID

Paul H. Jung ⓘ http://orcid.org/0000-0002-7267-0877

References

Alexander, C.H., 2002. Still rolling: Leslie Kish's "rolling samples" and the American Community Survey. *Survey Methodology*, 28 (1), 35–41.

Anselin, L., 1995. Local indicators of spatial association—LISA. *Geographical Analysis*, 27 (2), 93–115. doi:10.1111/j.1538-4632.1995.tb00338.x

Anselin, L., Syabri, I., and Kho, Y., 2006. GeoDa: an introduction to spatial data analysis. *Geographical Analysis*, 38 (1), 5–22. doi:10.1111/gean.2006.38.issue-1

Assunção, R.M. and Reis, E.A., 1999. A new proposal to adjust Moran's I for population density. *Statistics in Medicine*, 18, 2147–2162. doi:10.1002/(ISSN)1097-0258

Assunção, R.M., *et al.*, 2005. Empirical Bayes estimation of demographic schedules for small areas. *Demography*, 42 (3), 537–558.

Bazuin, J.T. and Fraser, J.C., 2013. How the ACS gets it wrong: the story of the American Community Survey and a small, inner city neighborhood. *Applied Geography*, 45, 292–302. doi:10.1016/j.apgeog.2013.08.013

Bradley, J.R., Wikle, C.K., and Holan, S.H., 2016. Bayesian spatial change of support for count-valued survey data with application to the American Community Survey. *Journal of the American Statistical Association*, 111 (514), 472–487. doi:10.1080/01621459.2015.1117471

Cliff, A.D. and Ord, J.K., 1981. *Spatial processes: models and applications*. London: Pion.

Cressie, N., 1992. Smoothing regional maps using empirical Bayes predictors. *Geographical Analysis*, 24 (1), 75–95. doi:10.1111/j.1538-4632.1992.tb00253.x

Datta, G.S., *et al.*, 2011. Bayesian benchmarking with applications to small area estimation. *Test*, 20 (3), 574–588. doi:10.1007/s11749-010-0218-y

Dima, A.L., 2018. Scale validation in applied health research: tutorial for a 6-step R-based psycho-metrics protocol. *Health Psychology and Behavioral Medicine*, 6 (1), 136–161. doi:10.1080/21642850.2018.1472602

Erickson, T. and Whited, T.M., 2002. Two-step GMM estimation of the errors-in-variables model using high-order moments. *Econometric Theory*, 18, 776–799. doi:10.1017/S0266466602183101

Fay, R., 2007. Imbedding model-assisted estimation into ACS estimation. In: *Proceedings of the 2007 Joint Statistical Meetings,* Salt Lake City, UT. Alexandria, VA: American Statistical Association, 2946–2953.

Folch, D.C., *et al.*, 2016. Spatial variation in the quality of American Community Survey estimates. *Demography*, 53 (5), 1535–1554. doi:10.1007/s13524-016-0499-1

Frost, C. and Thompson, S.G., 2000. Correcting for regression dilution bias: comparison of methods for a single predictor variable. *Journal of the Royal Statistical Society: Series A*, 163 (2), 173–189. doi:10.1111/1467-985X.00164

Fuller, W.A., 1987. *Measurement error models*. New York: Wiley.

Gage, L., 2006. Comparison of census 2000 and American Community Survey 1999–2001 estimates: San Francisco and Tulare Counties, California. *Population Research and Policy Review*, 25 (3), 243–256. doi:10.1007/s11113-006-9005-6

Ghosh, M. and Rao, J.N.K., 1994. Small area estimation: an appraisal. *Statistical Science*, 9 (1), 55–93. doi:10.1214/ss/1177010647

Griffin, R.A. and Albright, K., 2015. *Variance issues related to generalized regression estimation in the housing unit weighting operation*. Washington, DC: U.S. Census Bureau. No. ACS15-RER-05.

Hausman, J., 2001. Mismeasured variables in econometric analysis: problems from the right and problems from the left. *Journal of Economic Perspectives*, 15 (4), 57–67. doi:10.1257/jep.15.4.57

Hough, G.C. and Swanson, D.A., 2004. *The 1999–2001 American Community Survey and the 2000 census data quality and data comparisons: Multnomah county, Oregon*. Portland, OR: Population Research Center, Portland State University.

Hutcheon, J.A., Chiolero, A., and Hanley, J.A., 2010. Random measurement error and regression dilution bias. *BMJ*, 340, c2289. doi:10.1136/bmj.c293

Jung, P.H., Thill, J.-C., and Issel, M., 2018. Spatial autocorrelation statistics of areal prevalence rates under high uncertainty in denominator data. *Geographical Analysis* [online]. doi:10.1111/gean.12177

Jurjevich, J.R., *et al.*, 2018. Navigating statistical uncertainty: how urban and regional planners understand and work with American Community Survey (ACS) data for guiding policy. *Journal of the American Planning Association*, 84 (2), 112–126. doi:10.1080/01944363.2018.1440182

Kafadar, K., 1996. Smoothing geographical data, particularly rates of disease. *Statistics in Medicine*, 15 (23), 2539–2560. doi:10.1002/(ISSN)1097-0258

Kish, L., 1990. Rolling samples and censuses. *Survey Methodology*, 16 (1), 63–71.

Logan, J., *et al.*, 2018. The uptick in income segregation: real trend or random sampling variance. *American Journal of Sociology*, 124 (1), 185–222. doi:10.1086/697528

Macdonald, H., 2006. The American Community Survey: warmer (more current), but fuzzier (less precise) than the decennial census. *Journal of the American Planning Association*, 72 (4), 491–503. doi:10.1080/01944360608976768

Marshall, R., 1991. Mapping disease and mortality rates using empirical Bayes estimators. *Applied Statistics*, 40 (2), 283–294. doi:10.2307/2347593

National Research Council, 2007. *Using the American community survey: benefits and challenges.* Washington, DC: National Academy Press.

Navarro, A., 2012. An introduction to ACS statistical methods and lessons learned. In *Measuring People in Place Conference.* Boulder, CO. Available from: https://cupc.colorado.edu/workshops/measuring_people_in_place/themes/theme1/navarro.pdf [Accessed 22 April 2018].

Porter, A.T., Wikle, C.K., and Holan, S.H., 2015. Small area estimation via multivariate Fay-Herriot models with latent spatial dependence. *Australian & New Zealand Journal of Statistics*, 57 (1), 15–29. doi:10.1111/anzs.12101

Rao, J.N.K., 2005. Inferential issues in small area estimation: some new developments. *Statistics in Transition*, 7 (3), 513–526.

Rao, J.N.K. and Molina, I., 2015. *Small area estimation.* 2nd ed. Hoboken, NJ: Wiley.

Salvo, J., Lobo, P., and Calabrese, T., 2004. *Small area data quality: a comparison of estimates 2000 census and the 1999–2001 ACS Bronx, New York Test Site. New York.* New York: Department of City Planning, Population Division.

Schennach, S.M., 2004. Nonparametric regression in the presence of dynamic heteroskedasticity. *Econometric Theory*, 20, 1046–1093. doi:10.1017/S0266466604206028

Smith, G.D. and Phillips, A.N., 1996. Inflation in epidemiology: "The proof and measurement of association between two things" revisited. *BMJ*, 312, 1659–1661. doi:10.1136/bmj.312.7047.1659

Spielman, S.E., Folch, D., and Nagle, N., 2014. Patterns and causes of uncertainty in the American Community Survey. *Applied Geography*, 46, 147–157. doi:10.1016/j.apgeog.2013.11.002

Spielman, S.E. and Folch, D.C., 2015. Reducing uncertainty in the American Community Survey through data-driven regionalization. *PLoS ONE*, 10 (2), 1–21. doi:10.1371/journal.pone.0115626

Spielman, S.E. and Singleton, A., 2015. Studying neighborhoods using uncertain data from the American Community Survey: a contextual approach. *Annals of the Association of American Geographers*, 105 (5), 1003–1025. doi:10.1080/00045608.2015.1052335

Starsinic, M.D., 2005. American Community Survey: improving reliability for small area estimates. In: *Proceedings of the Joint Statistical Meetings, ASA Section on Survey Research Methods. (CD-ROM).* Alexandria, VA: American Statistical Association, 3592–3599.

Sun, M., Wong, D.W., and Kronenfeld, B.J., 2015. A classification method for choropleth maps incorporating data reliability information. *Professional Geographer*, 67 (1), 72–83. doi:10.1080/00330124.2014.888627

Sun, M. and Wong, D.W.S., 2010. Incorporating data quality information in mapping American Community Survey data. *Cartography and Geographic Information Science*, 37 (4), 285–300. doi:10.1559/152304010793454363

Sun, M. and Wong, D.W.S., 2017. Spatial aggregation as a means to improve attribute reliability. *Computers, Environment and Urban Systems*, 65, 15–27. doi:10.1016/j.compenvurbsys.2017.04.007

U.S. Census Bureau, 2000. U.S. Census 2000 long form questionnaire. Available from: https://www.census.gov/dmd/www/pdf/d02p.pdf [Accessed 22 April 2018].

U.S. Census Bureau, 2002. *Census 2000 basics.* Washington, DC: U.S. Government Printing Office.

U.S. Census Bureau, 2004. *Meeting 21 st century demographic data needs — implementing the American Community Survey.* Washington, DC: U.S. Census Bureau.

U.S. Census Bureau, 2008. *A compass for understanding and using American Community Survey data.* Washington, DC: U.S. Government Printing Office.

Wei, R. and Grubesic, T.H., 2017. An alternative classification scheme for uncertain attribute mapping. *Professional Geographer*, 69 (4), 604–615. doi:10.1080/00330124.2017.1288573

Wei, R., Tong, D., and Phillips, J.M., 2017. An integrated classification scheme for mapping estimates and errors of estimation from the American Community Survey. *Computers, Environment and Urban Systems*, 63, 95–103. doi:10.1016/j.compenvurbsys.2016.04.003

Appendix

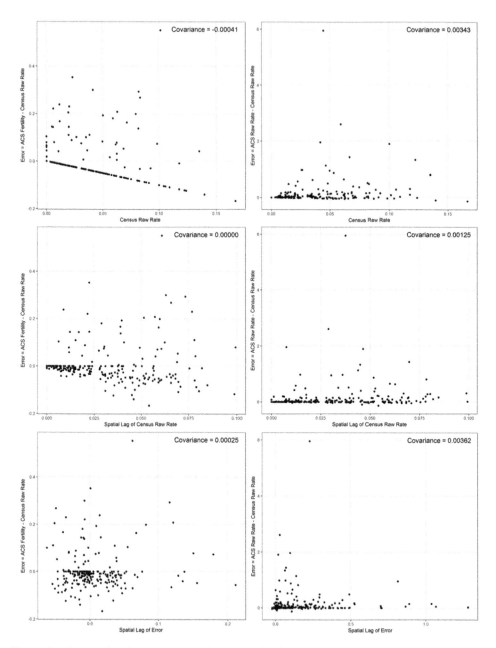

Figure A1. Scatter plots between error of ACS Teen birth rate and census raw rate and spatial lags.

Uncertainties in the geographic context of health behaviors: a study of substance users' exposure to psychosocial stress using GPS data

Mei-Po Kwan ⓘ, Jue Wang ⓘ, Matthew Tyburski, David H. Epstein, William J. Kowalczyk and Kenzie L. Preston

ABSTRACT

This study examined how contextual areas defined and operationalized differently may lead to different exposure estimates. Substance users' exposures to environmental stress (in terms of two variables: community social economic status and crime) were assessed from global positioning systems (GPS) data. Participants were 47 outpatients with substance use disorders admitted for methadone maintenance at a research clinic in Baltimore, Maryland. From 35.2 million GPS tracking points, we compared 7 different methods for defining activity space. The different methods yielded different exposure estimates, which would lead to different conclusions in studies using only one method. These results have important implications for future research on the effect of contextual influences on health behaviors and outcomes: whether a study observes any significant influence of an environmental factor on health may depend on what contextual units are used to assess individual exposure.

1. Introduction

Much of environmental health research is concerned with whether and how social and physical environments affect health (Berkman and Kawachi 2000; Diez-Roux 2001; Macintyre *et al.* 2002; Myers *et al.* 2016). To examine whether an environmental factor has a significant influence on a specific health behavior or outcome, it is first necessary to accurately measure individual exposure to the relevant environmental factor. An important issue in this measurement process is to identify the appropriate geographic area to be used as the contextual unit for deriving the values of various environmental variables (e.g. neighborhood socioeconomic status [SES]). Contextual areas in most studies in the past tend to be based on conventional delineations of people's residential neighborhood, which is often identified in terms of administrative units such as census tracts or postal code areas (Arcaya *et al.* 2016). Common assumptions underlying these conventional delineations of contextual areas include the neighborhood of residence is the most relevant area affecting health behaviors and outcomes, and neighborhood effects operate only through interactions among those residing within the same neighborhood unit. Further, individuals who live in the same areal unit are assumed to experience the same level of contextual influences, regardless of where they actually live within the area or how much time they spend outside their neighborhood of residence.

Using administrative units as the basis for evaluating neighborhood effects is convenient because they are often tied to data from censuses and surveys that can be used to derive contextual measures. However, people's exposures to influences from their social and physical environments are determined by where they go and how much time they spend there as they move around to undertake their daily activities (Cummins et al. 2007; Kwan 2009, 2013; Matthews 2008; Widener et al. 2013). People's activities (and thus environmental exposures) do not take place at one time point and wholly within any static, administratively bounded areal unit like census tracts or blocks. Residential location is one of the places where people spend their time, but for most people, the residential neighborhood does not capture many of their daily activities or the locations of these activities. In light of this, people's movement in space and time should be taken into account in order to more accurately estimate their environmental exposures and the effects of various contextual factors on their health behaviors and outcomes.

The methodological issue arising from the use of conventional contextual areas, largely in the form of static administrative areas, in health and geographic research has been recently articulated as the uncertain geographic context problem (UGCoP) (Kwan 2012). It refers to the problem that inferences about effects of area-based attributes (e.g. land-use mix) on individual behaviors or outcomes (e.g. physical activity) may be affected by how contextual areas or neighborhoods are geographically delineated. The problem 'arises because of the spatial uncertainty in the actual areas that exert contextual influences on the individuals being studied and the temporal uncertainty in the timing and duration in which individuals experienced these contextual influences' (Kwan 2012). As no researcher has complete and perfect knowledge of the 'true causally relevant' geographic context, all previous studies that used area-based contextual variables to explain individual behaviors or outcomes face the problem.

The UGCoP argument is strongly corroborated by findings from recent studies on environmental influences on body weight (James et al. 2014; Zhao et al. 2018), food access and its effects on health (Widener et al. 2013; Chen and Kwan 2015) and individual exposure to air pollution (Yoo et al. 2015; Dewulf et al. 2016; Park and Kwan 2017; Yu et al. 2018). These studies indicate that the UGCoP is a significant methodological problem in geographic and health research that seeks to understand how environmental factors affect health behaviors and outcomes. This paper seeks to contribute to this literature by showing how uncertainties in geographic context may affect exposure measures in a study of substance users' exposure to psychosocial stress using global positioning systems (GPS) data. It extends conventional concepts of neighborhood to a broader understanding of context based on people's activity spaces (i.e. the actual locations where people undertake their daily activities) (Golledge and Stimson 1997). It explores how context defined and operationalized differently may lead to different exposure estimates and different conclusions about their effects on health.

Specifically, the study examines substance users' exposure to psychosocial stress evaluated with two variables: a composite measure of community SES and crime. It uses a GPS dataset collected from outpatients with substance use disorders admitted for methadone maintenance at a research clinic in Baltimore, Maryland (Epstein *et al.* 2014). From 35.2 million GPS tracking points (GTPs), 7 contextual areas based on the notion of activity space are delineated for 47 participants. Measures of these participants' exposure to psychosocial stress are derived with these seven contextual areas, which are GTPs, GPS trajectory buffers (GTBs), standard deviational ellipses (SDEs) with one or two standard deviation(s) [SD(s)] (SDE1, SDE2), minimum convex polygons (MCPs), kernel density surfaces (KDSs) and home buffers (HBs). The results indicate that different delineations of contextual areas yield different exposure estimates, which vary considerably across the participants and exposure measures. These results have important implications for future research on the effect of contextual influences on health behaviors and outcomes: whether a study observes any significant influence of an environmental factor on health may depend on what contextual units are used to assess individual exposure.

2. Environmental influences on substance use behavior

Substance use and addiction are public health concerns that have wide-ranging social consequences. Considerable research has been conducted to identify risk and protective factors at the psychological and social levels (Boardman *et al.* 2001; Molina *et al.* 2012; Brenner *et al.* 2013). These factors generally fall into two broad categories: those that pertain to individuals and their interpersonal environments (e.g. attitudes and influences from peer groups) and those that reflect societal contextual conditions (e.g. neighborhood disorganization and economic deprivation).

Much of the early work in this area focused on individuals' own characteristics and those of their families and peers (Hawkins *et al.* 1992; Galea *et al.* 2004). Recent studies emphasize the additional impact of the neighborhood environment or context (which were largely derived from census-tract data) based on the hypothesis that neighborhoods can contribute to problem behavior (Sampson *et al.* 2002; Galea *et al.* 2005). For instance, researchers have observed associations between the physical and social environments and substance dependence (Kadushin *et al.* 1998) and between neighborhood disadvantage and the availability of illegal substances (Storr *et al.* 2004). Studies have also found that the association between neighborhood deprivation and substance use was partly mediated by differences in the level of psychological distress (Boardman *et al.* 2001). Many of these findings suggest that the influence of neighborhoods is causal. Specifically, neighborhood environment may significantly influence stress levels, which may influence substance use (Kwan *et al.* 2008; Mennis and Mason 2010, 2011; Brenner *et al.* 2013).

On the whole, however, research findings on the effect of neighborhood characteristics on illicit substance use are inconsistent. For instance, while several studies indicate that differences in neighborhood disadvantage – variously defined in terms of one or more characteristics based on census tracts, such as poverty status or an index capturing several community features – provide part of the explanation for higher levels of illicit substance use among adolescents and adults (Boardman *et al.* 2001; Hoffman 2002), others found that such neighborhood disadvantage measures are not significantly associated with adolescent substance use (e.g. Allison *et al.* 1999). Thus, among the many sociogeographic factors identified as relevant for illicit substance use and abuse, the role of neighborhood context remains less clear than those of individual, family and peer risk factors. While it is recognized that certain environmental or contextual conditions may invoke substance use among addicted individuals, there are significant conceptual and measurement issues in past studies that call for further investigation.

One important issue, as noted above, is that these studies have usually relied on exposure measurements based on people's residential neighborhoods, which are static and administratively bounded. Researchers have recently begun to adopt a dynamic notion of sociogeographic context (Kwan *et al.* 2008; Kwan 2009; Mason and Korpela 2009; Mennis and Mason 2011; Mennis *et al.* 2016; Epstein *et al.* 2014) and to use new data collection methods and analytical techniques (e.g. GPS data that tie to real-time ecological momentary assessment [EMA] data and geographic information systems [GIS]) (e.g. Mason *et al.* 2009; Mason and Korpela 2009).

This study uses a GPS dataset collected from substance users in Baltimore, Maryland, to explore how estimates of substance users' exposure to psychosocial stress may be affected by the contextual areas used to derive the exposure measures. It examines exposure to psychosocial stress using two variables: a composite measure of community SES and crime. The study seeks to shed light on the inconsistencies in previous findings and to contribute to the nascent literature on the UGCoP in health and GIScience research (e.g. James *et al.* 2014; Park and Kwan 2017; Shafran-Nathan *et al.* 2017; Helbich 2018; Wei *et al.* 2018; Zhao *et al.* 2018).

3. Data and methods

3.1. *Data collection and preprocessing*

The GPS dataset used in this study was collected in Baltimore, Maryland, from 2009 to 2011, using methods similar to those we described in a published pilot study with data collected in 2008–2009 (Epstein *et al.* 2014), but in a larger sample (Preston *et al.* 2018). The purpose of the project was to assess the associations between participants' real-time environmental psychosocial stress exposure and their substance use behavior. Participants were outpatients admitted for methadone maintenance at a research clinic in Baltimore. Eligibility criteria for enrollment were aged 18–65, physical dependence on opioids and evidence of cocaine and opiate use (self-report and urine). Exclusion criteria were any psychotic disorder listed in the *Diagnostic and Statistical Manual of Mental Disorders*, 4th Edition (DSM-IV), history of bipolar disorder or current major depressive disorder; current DSM-IV dependence on alcohol or sedative-hypnotics; cognitive impairment severe enough to preclude informed consent or valid self-report; and medical illness that would compromise participation. The Institutional Review Board of the National Institute on Drug Abuse approved the study, and participants gave written informed consent before enrollment. All data were covered by a Federal Certificate of Confidentiality.

Each participant carried a small GPS logger at all times during the study period. The GPS devices were set to log geolocation (latitude, longitude and altitude) every 20 m or every 15 min, whichever came first. This means that the GPS tracking data were not collected evenly in space and time but rather selectively: a location was logged if a participant moved more than 20 m from the previously recorded location, or a location was logged every 15 min if the participant did not move more than 20 m from the previously recorded location within the period. Each participant also carried a PalmPilot PDA at all times and was trained to use it as an electronic diary (ED), which provided data for the EMA component of the project. An ED entry was initiated whenever the participant (1) used any substance (e.g. cocaine, heroin or other opioid, marijuana, an amphetamine, benzodiazepines or alcohol) outside of a medical context or (2) felt overwhelmed, anxious or stressed more than usual. In addition to these event-triggered entries, participants responded to randomly timed prompts via their PDAs three times per day to report their mood, stress level, environmental setting, activities and degree of substance craving.

Analyses reported in this paper are based on a subsample of 47 participants from the larger multi-week project because of the computational intensity of the geospatial analysis involved. For these 47 participants, there were 35.2 million GTPs and the average tracking period is 107 days. Note that the tracking periods among these participants varied considerably: two participants had the shortest tracking period of 19 days, while two other participants had very long tracking periods that exceeded 200 days. Further, although the participants were outpatients admitted for methadone maintenance at a research clinic in the city of Baltimore, some of them spent considerable time outside the city. We thus expanded the study area to include the entire state of Maryland, which covers more than 90% of the total GPS records in the subsample.

Before data analysis was performed, a Python program was implemented to ensure consistency and eliminate errors in the GPS data. Each GPS track was checked to ensure proper temporal order and presence of at least two GPS points. Duplicate points were eliminated. Each track was also checked to make sure it met the criterion that a location was logged if a participant moved more than 20 m from the previously logged location, or every 15 min otherwise. Further, to reduce computational complexity, the duration between any pair of consecutive GPS records was assigned to the first point if a participant did not move more than 20 m during the period. This simplification is acceptable since 20 m is a small distance relative to the area of the contextual units delineated in the study.

3.2. *Contextual variables*

Past health studies have identified causal links between contextual influences and people's health behaviors. Particularly salient for substance use are social cohesion, social capital and collective efficacy, each of which can be protective against substance-related problems (Boardman *et al*. 2001; Latkin *et al*. 2005; Brenner *et al*. 2013). Conversely, neighborhoods that are low in these dimensions and have high crime rates tend also to have high rates of substance-related problems. In this study, we used two constructs to capture potential contextual influences on the participants. The first construct is community SES; the second is crime. Each is useful for representing social cohesion, collective efficacy and social disorder (e.g. Stahler *et al*. 2007; Mennis and Mason 2011; Epstein *et al*. 2014).

Community SES was operationalized in this study as a composite socioeconomic index (CSI) based on the modified Darden–Kamel composite index developed by Darden *et al.* (2010). It uses nine variables from census data and assigns a higher score to communities with higher SES. The nine variables are the percentage of residents with university degrees, median household income, the percentage of managerial and professional positions, median value of dwelling, median gross rent of dwelling, the percentage of homeownership, the percentage of households with vehicle, the percentage of population below poverty (reverse scored) and unemployment rate (reverse scored). The CSI was calculated as the sum of Z scores of the nine variables by using the following formula:

$$CSI_i = \sum_{j=1}^{n} \frac{V_{ij} - V_{j\,MD}}{SD(V_{j\,MD})} \qquad (1)$$

where CSI_i is the composite SES index for census tract i; n is the number of variables for the index (it is 9 for the modified Darden–Kamel composite index), V_{ij} is the jth variable for a given census tract i, $V_{j\,MD}$ is the mean of the jth variable for all census tracts in the state of Maryland and $SD(V_{j\,MD})$ is the SD of jth variable for all census tracts in Maryland. Since the GPS data were collected from 2009 to 2011, the American Community Survey census tract data of 2010 were used in this study to calculate the CSI for each census tract in Maryland (this was the finest resolution available on the website of the US Census at the time when analyses for this paper were performed). Figure 1 shows the CSI scores for all census tracts in Maryland.

Crime data were obtained from the Baltimore Neighborhood Indicator Alliance website, which provides data on two relevant measures: rate of violent crime and rate of general crime, each taken from reports to the Baltimore Police Department. Violent crimes include murder and nonnegligent homicide, forcible rape, aggravated assault

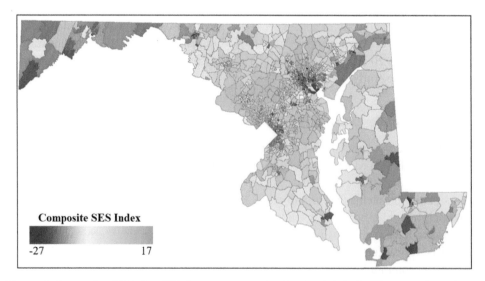

Figure 1. Composite SES index (CSI) for census tracts in Maryland. A high CSI score reflects better SES of a census tract and a low CSI score reflects lower SES of a census tract.

and robbery. General crime includes violent crime plus burglary, larceny-theft, motor vehicle theft and arson. Both violent and general crime rates are based on the number of relevant crimes per 100,000 residents in a particular community. This study used the broader category of general crime as the measure of crime. Since crime data at the level of census tract are available only for the city of Baltimore but not for the state of Maryland, analyses in this paper that used crime data focused only on Baltimore City.

3.3. Delineations of contextual areas and deriving exposure measures

To assess participants' exposure to psychosocial stress (measured by the CSI and crime), 7 delineations of individual contextual areas were implemented for the 47 selected participants – 6 using their GPS data, one using their home location. The six GPS-based delineations are GTPs, GTBs, SDEs with one or two SD(s) (SDE1, SDE2), KDSs and MCPs. The home-based delineations are HBs. Figure 2 illustrates five of these contextual areas for one participant (excluding GTP and HB to protect the participant's privacy). Appropriate methods of temporal or spatial interpolation were used to derive exposure measures from these contextual areas for each participant, and the contextual variables (CSI and crime) were standardized or normalized to ensure comparability among the participants. All data processing and analyses were performed with ArcGIS.

 1. *GTPs* – Participants' exposure to psychosocial stress was first assessed using their GTPs. Because the original GPS points are temporally uneven, a Python program was developed to interpolate data, using these steps: (1) The time difference between every pair of consecutive GPS points was calculated; (2) if the difference was longer than 1 s (say, *N* seconds), the program used linear interpolation to insert $N - 1$ more points evenly between the two. After this, each participant's GPS tracks had one point every

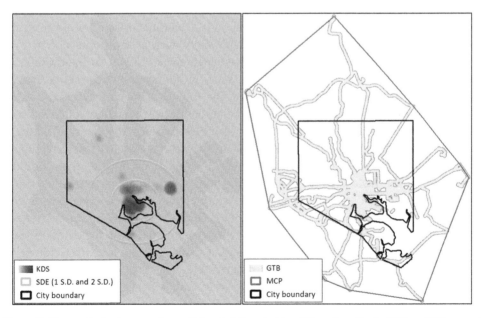

Figure 2. The contextual areas of a participant delineated by different methods (GTP and HB are not included for privacy concern).

second. Further, each of the contextual variables (CSI and crime) was computed as the sum of its values at all GPS points divided by the number of GPS points for each participant (this step standardized the contextual variable because the number of GPS points was different across participants).

2. GTBs – A GTB was created for each participant by covering the participant's GPS trajectories with a 200-m buffer area. Because this buffer area covers all the locations that a participant visited or passed during GPS logging, it can be used to assess participants' exposure to psychosocial stress in their daily life. This was done with a weighted sum of the contextual values based on the area of each census tract that was within a participant's GTB.

3 and 4. SDEs at one and two SD(s) (SDE1 and SDE2) – The SDE has been used in prior studies to delineate individual activity space and derive exposure measures (e.g. Sherman et al. 2005; Rainham et al. 2010). An SDE captures the geographic distribution and directional trend of a series of points and was derived in the following manner (Arcury et al. 2005; Wong and Lee 2005). First, the mean center of a participant's GPS points was derived. Then, the coordinates of each point in the set were transformed so that the center of the transformed coordinates became (0,0). The ellipse was finally obtained based on one or two SD(s) of the distances between each point and the transformed mean center along the rotated major and minor axes of the point set (Sherman et al. 2005). Since some past studies used one SD SDE while other studies used two, we derived both for comparison. For each participant, an SDE at one SD (SDE1) includes approximately 68% of the participant's GPS points within its boundary, while an SDE at two SDs (SDE2) contains approximately 95% (Sherman et al. 2005). To assess participants' exposure to psychosocial stress, a weighted sum of the values of a contextual variable (e.g. crime) was used, based on the area of each census tract that was within a participant's SDE.

5. Time-weighted KDSs – The KDS has also been used in prior studies to represent activity space and derive exposure measures. It is a density surface derived from the location of a set of points (and an associated weight such as the population at each point) using a kernel function and a predetermined search radius (or bandwidth). To generate a KDS for each participant in the study, a nonparametric kernel estimation method and a bandwidth of 1000 m was used. The 1000-m bandwidth avoids two issues associated with smaller bandwidths. First, the resulting density surface obtained with smaller bandwidths may approximate the GPS buffers, since the density surface is derived with only a few adjacent GPS points. Second, the resulting density surface obtained with smaller bandwidths may consist of many small discrete areas, because many cells in the raster surface will have 0 values, given that the smoothing kernel function covers only a small area and thus falls off rapidly. Based on some experiments with different bandwidth values, we found that 1000 m worked well for our dataset and study area. Using this bandwidth, the KDSs were generated based on the duration spent at each GTP as the weight on a raster layer with a spatial resolution of 100 × 100 m. To utilize the estimated kernel density values as the weight at each GTP for calculation of participants' contextual exposures, the raster cells with a density value larger than 0 were converted to points in a point-feature vector layer. The weighted sum of the values of a contextual variable (e.g. crime) was finally derived based on the kernel density value at each point.

6. MCPs – The MCP for a participant is the smallest convex polygon that contains all of the person's GTPs. To assess participants' exposure to psychosocial stress using the MCP, a weighted sum of the values of each contextual variable was derived, based on the area of each census tract that was within a participant's MCP.

7. HBs – Home location and residential neighborhood are the most commonly used contextual areas in past studies. Each participant's home location was used to construct a 1-km HB area for assessment of residential exposure to psychosocial stress. Note that information about participants' home location was not available in the dataset due to the need to protect their geoprivacy and data confidentiality. We inferred home locations based on the assumption that people tend to spend most of their time at or around their home location (e.g. sleep time and the time spent on home-based activities). This was achieved by using the total amount of time participants spent at and around their GTPs using the time-weighted KDS (which is a raster layer at the spatial resolution of 100 × 100 m): the location where a participant spent most time is treated as his/her home location. To assess participants' exposure to psychosocial stress using the home location buffer, a weighted sum of the values of each contextual variable was derived, based on the area of each census tract that was within a participant's home location buffer.

All seven contextual areas described above were implemented to derive measures of individual exposure to different levels of SES (CSI), but only the first six were used to evaluate exposure to crime. This was because the inferred home locations of many participants were outside Baltimore City, and crime data at the level of census tract were available only within the city.

4. Analysis and results

4.1. *Comparison of the size of different contextual areas*

We begin our analysis by comparing the size of 5 of the contextual area delineations described in the last section (Figure 3). The HBs had the same area for all participants because they were defined as 1-km buffer areas, and the GTPs method did not generate

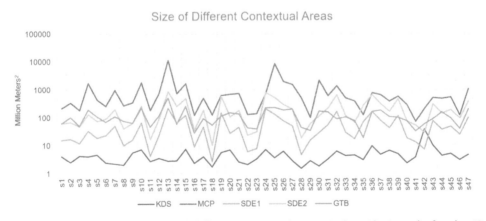

Figure 3. Comparison of the size of different contextual areas in logarithmic scale for the 47 participants. The vertical axis that indicates the size of contextual areas is on a logarithmic scale.

any area. Of the other five methods, four produced polygons with explicit boundaries (GTB, SDE1, SDE1, MCP) such that their areas were straightforward to calculate. For KDS (a continuous surface with different density values for different cells), the area was derived using a threshold density value that separates the top quantile from the four quantiles below it, and cells with values higher than the threshold were considered part of the contextual area.

The KDSs had sizes similar to the HBs, but with some variation across participants. The other four contextual areas varied considerably more across participants. MCP tended to be largest while SDE1 tended to be the smallest. While the sizes of these four contextual areas were largely different from each other, their differences among the participants tended to correlate (i.e. a participant with a small SDE1 would also have a small MCP, and vice versa). It is not surprising that the sizes of the SDE1 and SDE2 were highly correlated, because they were generated with the same method. However, the correlation between areas of MCP and GTB was somewhat surprising. It may reflect the fact that both methods generated polygons that included all GTPs.

The Wilcoxon signed rank test was used to evaluate the differences in size among these five contextual areas. It is a nonparametric equivalent of the paired-sample t-test; it does not assume normality in the data and thus can be used when this assumption is violated. Table 1 shows the results: the sizes of the five contextual areas were significantly different from each other. Note that the sizes of these five contextual areas varied considerably for some participants but much less for other participants. Participants with spatially dispersed GPS trajectories tended to have MCPs that were much larger than other delineations of contextual areas (and thus also tended to have the largest variation among the five contextual areas). Conversely, participants with relatively circumscribed GPS trajectories had smaller and similarly sized contextual areas. This could have practical implications, because the size of the contextual areas can affect estimates of exposure (e.g. to crime), as we discuss in the next section.

4.2. Comparison of individual exposure to community SES

The community SES index (CSI) described earlier was used as a surrogate to assess participants' exposure to the psychosocial stress presumed to accompany social disorder. We used all of the 7 contextual areas to derive 7 CSIs for each of the 47 participants. As shown in Figure 4, values of CSI in the participants' activity spaces varied from −23 to

Table 1. Wilcoxon signed rank correlation test of the size of different contextual areas.

Measurement pair	Z	Asymp. sig. (two-tailed)
MCP–KDS	−5.968[a]	<0.001
SDE1–KDS	−5.714[a]	<0.001
SDE2–KDS	−5.947[a]	<0.001
GTB–KDS	−5.968[a]	<0.001
SDE1–MCP	−5.968[b]	<0.001
SDE2–MCP	−5.767[b]	<0.001
GTB–MCP	−5.968[b]	<0.001
SDE2–SDE1	−5.968[a]	<0.001
GTB–SDE1	−5.259[a]	<0.001
GTB–SDE2	−3.037[b]	0.002

[a]Based on negative ranks.
[b]Based on positive ranks.

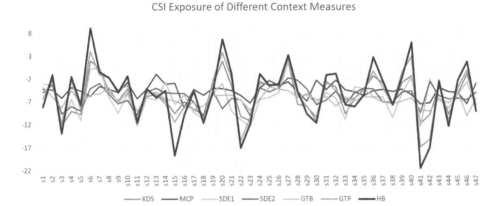

Figure 4. Comparison of CSI exposures based on different contextual areas for the 47 participants.

10, where a higher CSI score indicates higher SES, and vice versa. Variation among participants was largest with the HBs (SD = 6.54) and the GTPs (SD = 4.91), and smallest with MCP (SD = 1.55) and GTB (SD = 1.52). This is perhaps because delineations other than the HBs and the GTP are relatively large polygons that cover many census tracts, attenuating the influence of extremes.

To explore how values of the CSI varied within participants, we examined variations in the SD of the CSI in Figure 5. The figure shows that the CSIs obtained with the seven contextual areas for three participants (s6, s20 and s42) had very large SDs, while the CSI for three other participants (s12, s26 and s35) had very small SDs. We further investigated these six participants using geovisualizations, whose results are discussed here but cannot be shown due to IRB requirements to protect participants' privacy. All three participants with large CSI SDs had relatively small activity spaces with fairly even coverage within those spaces. On the other hand, two out of the three participants with small CSI SDs had relatively large activity spaces with strong directional trends. As suggested earlier, larger contextual areas cover many census tracts and the CSI derived with them tend to vary less.

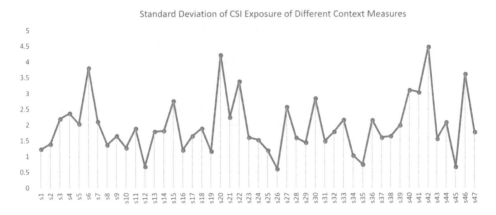

Figure 5. Standard deviation of CSI exposures based on different contextual areas for the 47 participants.

These results indicate that variation in CSI exposure can vary depending on the method used to define contextual areas – not only across participants but also for the same participant. A Wilcoxon signed rank test (Table 2) indicated that 6 out of 15 pairs of CSI estimates were significantly different from each other: MCP versus KDS, SDE1 versus KDS, SDE1 versus MCP, SDE2 versus MCP, GTB versus MCP and GTP versus SDE1. Note that four of these six pairs involve MCP and three of them involve SDE. Also note that all of the six pairs involving the HBs were not significantly different, which means that the CSI derived with the HBs were not significantly different from the CSI assessed using the other six contextual areas. This result may be due to the fact that the HBs were defined around the locations where participants spent most of their time, and participants' home location also figured prominently in the definitions of the other contextual areas. For instance, the GTPs, GPS buffers (GTB), SDEs (SDE1 and SDE2) and time-weighted KDSs tended to be skewed toward the place where participants spent the most time (i.e. their home location) because this place was heavily weighted in the data set.

4.3. Comparison of individual exposure to crime

Figure 6 shows estimates of general crime exposures as assessed by six of the contextual areas (HBs were not included in this analysis because the inferred home locations of many participants were outside Baltimore City, and crime data at the level of census tract were available only within the city). Exposures assessed with KDS (SD = 19.43) and GTP (SD = 26.17) had the largest variations among the 47 participants; exposures assessed with MCP (SD = 5.88) had the least variation.

Wilcoxon signed rank tests (Table 3) showed that 11 out of the 15 pairs of crime-exposure estimates were different from each other: MCP versus KDS, SDE1 versus KDS, SDE2 versus KDS, SDE1 versus MCP, SDE2 versus MCP, GTB versus MCP, GTP versus MCP, GTB versus SDE1, GTP versus SDE1, GTB versus SDE2 and GTP versus SDE2. Exposures for only four pairs of contextual areas were not significantly different: GTB versus KDS, GTP versus KDS, GTP versus GTB and SDE2 versus SDE1.

Table 2. Wilcoxon signed rank correlation test of CSI exposures based on different contextual areas.

Measurement pair	Z	Asymp. sig. (two-tailed)
MCP–KDS	−2.085[a]	0.037
SDE1–KDS	−2.032[b]	0.042
SDE2–KDS	0.000[c]	1.000
GTB–KDS	−0.995[b]	0.320
GTP–KDS	−1.884[a]	0.060
SDE1–MCP	−3.026[b]	0.002
SDE2–MCP	−3.418[b]	0.001
GTB–MCP	−5.492[b]	<0.001
GTP–MCP	−1.058[b]	0.290
SDE2–SDE1	−1.545[a]	0.122
GTB–SDE1	−0.296[a]	0.767
GTP–SDE1	−2.032[a]	0.042
GTB–SDE2	−1.778[b]	0.075
GTP–SDE2	−1.090[a]	0.276
GTP–GTB	−1.640[a]	0.101

[a]Based on negative ranks.
[b]Based on positive ranks.
[c]The sum of negative ranks equals the sum of positive ranks.

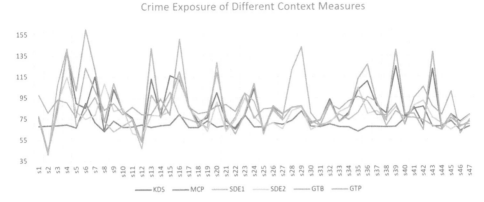

Figure 6. Comparison of crime exposures based on different contextual areas for the 47 participants.

Table 3. Wilcoxon signed rank correlation test of crime exposures based on different contextual areas.

Measurement pair	Z	Asymp. Sig. (two-tailed)
MCP–KDS	−4.635[a]	<0.001
SDE1–KDS	−3.005[a]	0.003
SDE2–KDS	−2.730[a]	0.006
GTB–KDS	−1.217[b]	0.224
GTP–KDS	−1.355[b]	0.176
SDE1–MCP	−3.365[b]	0.001
SDE2–MCP	−3.831[b]	<0.001
GTB–MCP	−5.894[b]	<0.001
GTP–MCP	−4.561[b]	<0.001
SDE2–SDE1	−0.224[a]	0.823
GTB–SDE1	−3.291[b]	0.001
GTP–SDE1	−2.794[b]	0.005
GTB–SDE2	−3.683[b]	<0.001
GTP–SDE2	−2.593[b]	0.010
GTP–GTB	−0.349[a]	0.727

[a]Based on positive ranks.
[b]Based on negative ranks.

Figure 7 shows the SDs of general crime exposures assessed by the 6 contextual areas for the 47 participants. Three participants (s4, s6 and s16) had very high SDs; two participants (s31 and s40) had very low SDs. Geovisualizations did not identify any distinctive patterns that would help explain these differences across participants. The differences may have occurred because crime exposures are highly dependent on the unique geographic distribution of crime in the study area and the exposure measures obtained are not heavily influenced by the spatial configurations of the participants' activity spaces. This in turn may be due to fact that contextual areas other than the GTP and GTB are relatively large polygons that cover many census tracts, attenuating the influence of extremes and thus producing similarly moderate estimates for crime exposure.

4.4. *Factor analysis of the associations among the contextual variables*

To assess the extent to which contextual variables derived with different contextual areas capture similar or different information regarding exposures, their associations were examined

Figure 7. Standard deviation of crime exposures based on different contextual areas for the 47 participants.

Table 4. Factor loadings of CSI exposures based on different contextual areas.

	Factor loadings			Communalities
	Factor 1	Factor 2	Factor 3	
KDS	**0.896**	0.351	0.155	0.951
MCP	0.093	0.054	**0.920**	0.858
SDE1	0.580	**0.743**	0.083	0.896
SDE2	0.183	**0.927**	0.230	0.946
GTB	0.291	0.292	**0.770**	0.762
GTP	**0.933**	0.162	0.256	0.962
Variance explained	2.136	1.650	1.588	5.374
% of Variance	35.61	27.50	26.46	89.57

Note: Loadings over 0.7 are in bold type.

using factor analysis. Three factors were extracted from the six measures, and a varimax rotation was performed to make these factors more interpretable. These three rotated factors together explained 89.57% of the total variance for CSI exposures (Table 4) and 87.02% of the total variance for crime exposures (Table 5). It can be seen from these tables that all six exposure measures had large communalities, which indicate that a large amount of their variance has been extracted. Figures 8 and 9 visualize the factor loadings of the six measures with three-dimensional scatterplots, in which clusters of different types of exposure measures can be easily identified. As these figures show, KDS and GTP measures had high loadings (over 0.7) on Factor 1, on which SDE1 also had moderate loadings (over 0.5). Furthermore, SDE1 and

Table 5. Factor loadings of crime exposures based on different contextual areas.

	Factor loadings			Communalities
	Factor 1	Factor 2	Factor 3	
KDS	**0.899**	0.011	0.351	0.932
MCP	0.036	**0.926**	−0.048	0.861
SDE1	0.508	−0.012	0.698	0.745
SDE2	0.212	0.198	**0.900**	0.895
GTB	0.121	**0.869**	0.275	0.845
GTP	**0.937**	0.175	0.185	0.943
Variance explained	2.005	1.682	1.533	5.220
% of Variance	33.42	28.04	25.56	87.02

Note: Loadings over 0.7 are in bold type.

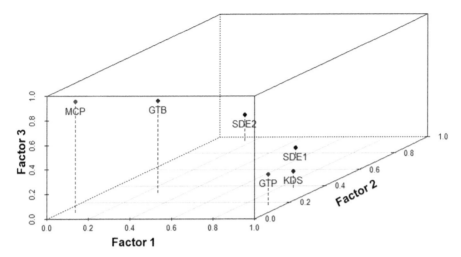

Figure 8. Factor loadings of CSI exposures based on different contextual areas on Factors 1, 2 and 3.

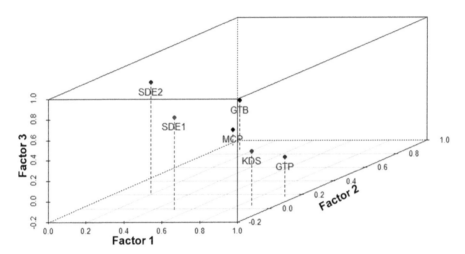

Figure 9. Factor loadings of crime exposures based on different contextual areas on Factors 1, 2 and 3.

SDE2 both had high loadings (over 0.7) on Factor 2, while MCP and GTB had high loadings (over 0.7) on Factor 3. As to the factor loadings for crime exposure, similar patterns were found. KDS and GTP measures had high loadings (over 0.7) on Factor 1, while MCP and GTB had high loadings (over 0.7) on Factor 2. In addition, only SDE2 had high loadings (over 0.7) on Factor 3, on which SDE1 also had moderate loadings (over 0.5). Overall, a general pattern can be identified: values of exposure measures obtained with KDS and GTP were similar, those obtained with SDE1 and SDE2 fell into another group and those obtained with MCP and GTB fell into a third group.

4.5. *Regression analyses on CSI and crime estimates as predictors of substance use behavior*

Multivariate linear regression models were used to examine the associations between the participants' exposure to psychosocial stress (based on the CSI and crime) and their substance use behaviors. Multilevel models were not feasible because there was no obvious spatial clustering in participants' home location and the number of participants was small. Further, given that the average GPS tracking period was only 107 days, both community SES and crime in the study area were assumed to be stable enough not to require longitudinal analyses.

The dependent variables were two measures of substance use for each participant: total and standardized. The total substance use score for each participant was the sum of seven substance use indicators based on urine tests performed three times each week during the study period. Each indicator reflects the presence or absence of amphetamine, barbiturates, benzodiazepines, opiates, cocaine, phencyclidine and cannabinoids (1 for presence and 0 for absence). Because the number of total urine tests differed among participants, a standardized substance use score for each participant was obtained by dividing the total score by the number of urine tests. Both substance use scores were log-transformed to minimize their skewness. The main independent variables in all regression models were the CSI and crime exposure measured by the seven contextual areas. There were thus 14 models: 7 used the total substance use score, and 7 used the standardized substance use score. Another 14 models included participants' demographic characteristics: gender, race, income, marital status and employment status. There were no significant associations among substance use, contextual exposures and demographic characteristics, and R^2 values of the regression models were uniformly low, ranging from 0.020 to 0.337.

However, the finding that exposures to CSI and crime were not significantly associated with substance use behavior does not necessarily mean that these contextual influences are unimportant. All participants were substance users admitted for methadone maintenance at a research clinic, and there was no control group of nonusers, so it is possible that participants' substance use behaviors were influenced more by the momentary psychosocial stress they experienced at particular places and times (e.g. sighting of law enforcement personnel at a particular time and place) than by the general sociogeographic contexts assessed using their activity spaces (Kwan 2018).

5. Conclusion

This study used a GPS dataset collected from substance users in Baltimore, Maryland, to explore how estimates of substance users' exposure to psychosocial stress may be affected by the contextual areas used to derive the exposure measures. It examined how geographic context defined and operationalized differently may lead to different exposure estimates and different conclusions about their effects on health, using two variables (a composite measure of community SES and crime) to assess participants' exposure to psychosocial stress. The study sought to shed light on the inconsistencies in previous findings and contribute to the nascent literature on the UGCoP in health and GIScience research.

The results indicate that different delineations of contextual areas yield different exposure estimates, and this may lead to different conclusions. For instance, values of CSI based on different contextual areas varied greatly, and variation among participants was largest with the HBs and the GTPs, and smallest with MCP and GTB. Six out of 15 pairs of CSI estimates were significantly different. With respect to crime, exposures assessed with KDS and GTP had the largest variations among the 47 participants, and exposures assessed with MCP had the least variation. Eleven out of 15 pairs of crime-exposure estimates were different from each other. These results suggest that contextual uncertainty is an important issue when examining the effects of various contextual factors on substance use behaviors. It has important implications for future research on the effect of contextual influences on health behaviors and outcomes: whether a study observes any significant influence of an environmental factor on health depends on what contextual units are used to assess individual exposure.

Several areas seem promising for future research. First, as pointed out earlier in this paper, participants' substance use behaviors may be influenced more by the momentary psychosocial stress they experienced at particular places and times than by the socio-geographic contexts assessed using their activity spaces (Kwan 2018). This may be an important basis for momentary interventions. Second, as our ongoing work indicates, whether a participant used a substance may be influenced by the sociogeographic context he/she experienced several hours ago; substance use may be a response to psychosocial stress with a certain time lag. Future studies should try to examine the particular temporality of the association between contextual influences and people's substance use behaviors. Lastly, individual response to the same contextual influences may be highly idiosyncratic (Kwan 2018). A substance user may feel stressed in places where nonusers typically feel safe (e.g. areas with high SES, or low-crime areas that are heavily patrolled by police). Future studies should try to take into account the idiosyncratic responses of individuals to the same contextual influences.

Acknowledgments

The authors thank the reviewers for their helpful comments. They had sole responsibility for the design, analysis and interpretation of the results of the study.

Disclosure statement

No potential conflict of interest was reported by the authors.

Funding

This research was supported by the University of Illinois Research Board and the Intramural Research Program (IRP) of the National Institute on Drug Abuse (NIDA), NIH. In addition, Mei-Po Kwan was supported by a John Simon Guggenheim Memorial Foundation Fellowship while conducting this research.

ORCID

Mei-Po Kwan ⓘ http://orcid.org/0000-0001-8602-9258
Jue Wang ⓘ http://orcid.org/0000-0002-6305-4298

References

Allison, K.W., *et al.* 1999. Adolescent substance use: preliminary examinations of school and neighborhood context. *American Journal of Community Psychology*, 27 (2), 111–141. doi:10.1023/A:1022879500217

Arcaya, M.C., *et al.*, 2016. Research on neighborhood effects on health in the United States: a systematic review of study characteristics. *Social Science & Medicine*, 168, 16–29. doi:10.1016/j.socscimed.2016.08.047

Arcury, T.A., *et al.* 2005. The effects of geography and spatial behavior on health care utilization among the residents of rural region. *Health Services Research*, 40 (1), 135–155. doi:10.1111/j.1475-6773.2005.00346.x

Berkman, L. and Kawachi, I., 2000. *Social epidemiology*. New York: Oxford University Press.

Boardman, J., *et al.* 2001. Neighborhood disadvantage, stress, and drug use among adults. *Journal of Health and Social Behavior*, 42 (2), 151–165. doi:10.2307/3090175

Brenner, A., *et al.*, 2013. Neighborhood context and perceptions of stress over time: an ecological model of neighborhood stressors and interpersonal and intrapersonal resources. *American Journal of Community Psychology*, 51, 544–556. doi:10.1007/s10464-013-9571-9

Chen, X. and Kwan, M.-P., 2015. Contextual uncertainties, human mobility, and perceived food environment: the uncertain geographic context problem in food access research. *American Journal of Public Health*, 105 (9), 1734–1737. doi:10.2105/AJPH.2013.301857

Cummins, S., *et al.*, 2007. Understanding and representing 'place' in health research: A relational approach. *Social Science & Medicine*, 65, 1825–1838. doi:10.1016/j.socscimed.2007.05.036

Darden, J., *et al.* 2010. The measurement of neighborhood socioeconomic characteristics and black and white residential segregation in metropolitan Detroit: implications for the study of social disparities in health. *Annals of the Association of American Geographers*, 100 (1), 137–158. doi:10.1080/00045600903379042

Dewulf, B., *et al.*, 2016. Dynamic assessment of exposure to air pollution using mobile phone data. *International Journal of Health Geographics*, 15, 14. doi:10.1186/s12942-016-0042-z

Diez-Roux, A.V., 2001. Investigating neighborhood and area effects on health. *American Journal of Public Health*, 91, 1783e1789.

Epstein, D.H., *et al.*, 2014. Real-time tracking of neighborhood surroundings and mood in urban drug misusers: application of a new method to study behavior in its geographical context. *Drug and Alcohol Dependence*, 134, 22–29. doi:10.1016/j.drugalcdep.2013.09.007

Galea, S., Nandi, A., and Vlahov, D., 2004. The social epidemiology of substance use. *Epidemiologic Reviews*, 26, 36–52. doi:10.1093/epirev/mxh007

Galea, S., Rudenstine, S., and Vlahov, D., 2005. Drug use, misuse, and the urban environment. *Drug and Alcohol Review*, 24, 127–136. doi:10.1080/09595230500102509

Golledge, R.G. and Stimson, R.J., 1997. *Spatial behavior: A geographic perspective*. New York: Guilford.

Hawkins, J.D., Catalano, R.F., and Miller, J.Y., 1992. Risk and protective factors for alcohol and other drug problems in adolescence and early adulthood: implications for substance abuse prevention. *Psychological Bulletin*, 112 (1), 64–105. doi:10.1037/0033-2909.112.1.64

Helbich, M., 2018. Toward dynamic urban environmental exposure assessments in mental health research. *Environmental Research*, 161, 129–135. doi:10.1016/j.envres.2017.11.006

Hoffmann, J.P., 2002. The community context of family structure and adolescent drug use. *Journal of Marriage and Family*, 64, 314–330. doi:10.1111/j.1741-3737.2002.00314.x

James, P., *et al.*, 2014. Effects of buffer size and shape on associations between the built environment and energy balance. *Health & Place*, 27, 162–170. doi:10.1016/j.healthplace.2014.02.003

Kadushin, C., *et al.*, 1998. The substance use system: social and neighborhood environments associated with substance use and misuse. *Substance Use and Misuse*, 33 (8), 1681–1710.

Kwan, M.P., *et al.* 2008. Reconceptualizing sociogeographic context for the study of drug use, abuse, and addiction. *In*: Y.F. Thomas, D. Richardson, and I. Cheung, eds. *Geography and drug addiction*. Berlin: Springer, 437–446.

Kwan, M.-P., 2009. From place-based to people-based exposure measures. *Social Science and Medicine*, 69 (9), 1311–1313. doi:10.1016/j.socscimed.2009.07.013

Kwan, M.-P., 2012. The uncertain geographic context problem. *Annals of the Association of American Geographers*, 102 (5), 958–968. doi:10.1080/00045608.2012.687349

Kwan, M.-P., 2013. Beyond space (as we knew it): toward temporally integrated geographies of segregation, health, and accessibility. *Annals of the Association of American Geographers*, 103 (5), 1078–1086. doi:10.1080/00045608.2013.792177

Kwan, M.-P., 2018. The limits of the neighborhood effect: contextual uncertainties in geographic, environmental health, and social science research. *Annals of the American Association of Geographers*, 108 (6).

Latkin, C.A., *et al.* 2005. Neighborhood social disorder as a determinant of drug injection behaviors: a structural equation modeling approach. *Health Psychology*, 24 (1), 96–100. doi:10.1037/0278-6133.24.1.96

Macintyre, S., Ellaway, A., and Cummins, S., 2002. Place effects on health: how can we conceptualise, operationalise and measure them? *Social Science & Medicine*, 55, 125e139. doi:10.1016/S0277-9536(01)00214-3

Mason, M.J., Cheung, I., and Walker, L., 2009. Creating a geospatial database of risks and resources to explore urban adolescent substance use. *Journal of Prevention & Intervention in the Community*, 37, 21–34. doi:10.1080/10852350802498391

Mason, M.J. and Korpela, K., 2009. Activity spaces and urban adolescent substance use and emotional health. *Journal of Adolescence*, 32 (4), 925–939. doi:10.1016/j.adolescence.2008.08.004

Matthews, S.A., 2008. The salience of neighborhood: some lessons from sociology. *American Journal of Preventive Medicine*, 34 (3), 257–259. doi:10.1016/j.amepre.2007.12.001

Mennis, J. and Mason, M.J., 2010. Social and geographic contexts of adolescent substance use: the moderating effects of age and gender. *Social Networks*, 34, 150–157.

Mennis, J. and Mason, M.J., 2011. People, places, and adolescent substance use: integrating activity space and social network data for analyzing health behavior. *Annals of the Association of American Geographers*, 101 (2), 272–291. doi:10.1080/00045608.2010.534712

Mennis, J., Stahler, G.J., and Mason, M.J., 2016. Risky substance use environments and addiction: a new frontier for environmental justice research. *International Journal of Environmental Research and Public Health*, 13 (6), 607. doi:10.3390/ijerph13121252

Molina, K., Alegria, M., and Chen, C., 2012. Neighborhood context and substance use disorders: A comparative analysis of racial and ethnic groups in the United States. *Drug and Alcohol Dependence*, 125, 35–43. doi:10.1016/j.drugalcdep.2012.05.027

Myers, C.A., Denstel, K.D., and Broyles, S.T., 2016. The context of context: examining the associations between healthy and unhealthy measures of neighborhood food, physical activity, and social environments. *Preventive Medicine*, 93, 21–26. doi:10.1016/j.ypmed.2016.09.009

Park, Y.M. and Kwan, M.-P., 2017. Individual exposure estimates may be erroneous when spatio-temporal variability of air pollution and human mobility are ignored. *Health & Place*, 43, 85–94. doi:10.1016/j.healthplace.2016.10.002

Preston, K.L., *et al.*, 2018. Exacerbated craving in the presence of stress and drug cues in drug-dependent patients. *Neuropsychopharmacology*, 43, 859–867. doi:10.1038/npp.2017.275

Rainham, D., *et al.*, 2010. Conceptualizing the healthscape: contributions of time geography, location technologies and spatial ecology to place and health research. *Social Science & Medicine*, 70, 668–676. doi:10.1016/j.socscimed.2009.10.035

Sampson, R.J., Morenoff, J.D., and Gannon-Rowley, T., 2002. Assessing neighborhood effects: social processes and new directions in research. *Annual Review of Sociology*, 28, 443–478. doi:10.1146/annurev.soc.28.110601.141114

Shafran-Nathan, R., *et al.*, 2017. Exposure estimation errors to nitrogen oxides on a population scale due to daytime activity away from home. *Science of the Total Environment*, 580, 1401–1409. doi:10.1016/j.scitotenv.2016.12.105

Sherman, J.E., *et al.*, 2005. A suite of methods for representing activity space in a healthcare accessibility study. *International Journal of Health Geographics*, 4, 24. doi:10.1186/1476-072X-4-24

Stahler, G., *et al.*, 2007. The effect of individual, program, and neighborhood variables on continuity of treatment among dually diagnosed individuals. *Drug and Alcohol Dependence*, 87, 54–62. doi:10.1016/j.drugalcdep.2006.07.010

Storr, C.L., *et al.* 2004. Unequal opportunity: neighbourhood disadvantage and the chance to buy illegal drugs. *Journal of Epidemiology and Community Health*, 58 (3), 231–237. doi:10.1136/jech.2003.007575

Wei, Q., *et al.*, 2018. Using individual GPS trajectories to explore foodscape exposure: A case study in Beijing metropolitan area. *International Journal of Environmental Research and Public Health*, 15, 405. doi:10.3390/ijerph15061188

Widener, M.J., *et al.*, 2013. Using urban commuting data to calculate a spatiotemporal accessibility measure for food environment studies. *Health & Place*, 21, 1–9. doi:10.1016/j.healthplace.2013.01.004

Wong, D.W.S. and Lee, J., 2005. *Statistical analysis of geographic information with ArcView GIS and ArcGIS*. New York: Wiley.

Yoo, E., *et al.* 2015. Geospatial estimation of individual exposure to air pollutants: moving from static monitoring to activity-based dynamic exposure assessment. *Annals of the Association of American Geographers*, 105 (5), 915–926. doi:10.1080/00045608.2015.1054253

Yu, H., *et al.*, 2018. Using cell phone location to assess misclassification errors in air pollution exposure estimation. *Environmental Pollution*, 233, 261–266. doi:10.1016/j.envpol.2017.10.077

Zhao, P., Kwan, M.-P., and Zhou, S., 2018. The uncertain geographic context problem in the analysis of the associations between obesity and the built environment in Guangzhou. *International Journal of Environmental Research and Public Health*, 15, 308. doi:10.3390/ijerph15061188

Exploring the uncertainty of activity zone detection using digital footprints with multi-scaled DBSCAN

Xinyi Liu ⓘD, Qunying Huang ⓘD and Song Gao ⓘD

ABSTRACT

The density-based spatial clustering of applications with noise (DBSCAN) method is often used to identify individual activity clusters (i.e., zones) using digital footprints captured from social networks. However, DBSCAN is sensitive to the two parameters, *eps* and *minpts*. This paper introduces an improved density-based clustering algorithm, Multi-Scaled DBSCAN (M-DBSCAN), to mitigate the detection uncertainty of clusters produced by DBSCAN at different scales of density and cluster size. M-DBSCAN iteratively calibrates suitable local *eps* and *minpts* values instead of using one global parameter setting as DBSCAN for detecting clusters of varying densities, and proves to be effective for detecting potential activity zones. Besides, M-DBSCAN can significantly reduce the noise ratio by identifying all points capturing the activities performed in each zone. Using the historic geo-tagged tweets of users in Washington, D.C. and in Madison, Wisconsin, the results reveal that: 1) M-DBSCAN can capture dispersed clusters with low density of points, and therefore detecting more activity zones for each user; 2) A value of 40 m or higher should be used for *eps* to reduce the possibility of collapsing distinctive activity zones; and 3) A value between 200 and 300 m is recommended for *eps* while using DBSCAN for detecting activity zones.

1. Introduction

Human mobility study is a significant research thrust in GIScience by contributing to various applications, such as examining human behaviors and mobility patterns, revealing underlying urban spatial structure and dynamics, or understanding the evolution of epidemics and spatial spread of diseases (Kang *et al.* 2012, Noulas *et al.* 2012a, Kwan 2013, Richardson *et al.* 2013, Gao 2015, Xu *et al.* 2016, Huang 2017). While surveys, GPS data, and mobile phone records have been primarily used to explore human movement behaviors and patterns, social media now is widely used as a new source that captures human regular activity patterns (e.g. commuting time) and population dynamics (Gao *et al.* 2014, Huang and Wong 2015, Steiger *et al.* 2015, Luo *et al.* 2016). However, digital footprints recorded by social media and represented as a series of spatiotemporal (ST) points are highly sparse and irregular in both spatial and temporal dimensions, and thus various spatial data processing and data mining methods have to be applied before meaningful mobility patterns can be discovered. Herein, spatial clustering is one of the important methods to make sense of digital footprints by grouping the sparse and irregular ST points to clusters (Mennis and Guo 2009). During the work of modeling, visualizing and mining individual digital footprints captured through social media, it is a paramount preprocessing component for various data mining algorithms and models, such as home location reference and user future location prediction models (Mahmud et al. 2012, Mathew et al. 2012, Lin and Cromley 2018), to discover the regular activity zones and points of interest (POIs) that an individual or a group of users with similar movement behaviors regularly visit or stay.

Among all spatial clustering algorithms, the density-based spatial clustering of applications with noise (DBSCAN; Ester et al. 1996) is very popular with the capability of detecting clusters of arbitrary shape with noise, and only needs to supply two input parameters: minpts, and eps; where eps is the search radius of a point, i.e. the neighborhood of a point, and minpts is the minimum number of points that the neighborhood should include to form a cluster (Moreira et al. 2013). In DBSCAN, points with a neighborhood more than minpts are categorized as core points, and points reachable by a core point are border points of the cluster. Points not within the eps search radius of any core point are treated as noise. Both core points and border points are considered as the clustered points.

At present, a few studies have provided some general guidelines on setting up a minpts and an eps value for various datasets (Ester et al. 1996, Zhou et al. 2004). For example, Ester et al. (1996) reveal that using a minpts value <4 may capture noisy points (outliers) in clusters and the clustering results most likely are similar with a minpts value ≥4. While clustering dense GPS trajectories, Zhou et al. (2004) use an eps value of 20 meters (m), approximating to the uncertainty in GPS positioning. With regard to geo-tagged social media data clustering analysis, Hu et al. (2015) compare the DBSCAN results with 25 different parameter combinations and demonstrate that eps = 200 m search radius can help identify the well-known urban areas of interest neighborhoods in cities using geo-tagged Flickr photos. However, the activity zone and trajectory detection is highly uncertain and sensitive to the DBSCAN parameter values with sparse trajectory points collected from social networks: while the clustering results produced by smaller minpts and/or eps values likely collapse footprint points in one activity zone into smaller clusters and capture clusters of random activities, the derived trajectories are likely more precise in space; the results generated by larger minpts and/or eps values include fewer but larger clusters, more likely merging footprint points in different activity zones as one cluster. Correspondingly, the resulting trajectories are likely less precise in space.

In addition, while DBSCAN is able to identify clusters of arbitrary shapes, it is insufficient to deal with data with clusters of different densities (or neighborhoods) as it fails to detect the core points of varying density clusters with a single eps value (Ertöz et al. 2003). For example, the points in the bottom left cluster (Figure 1) are more compacted or denser in space and with 25 as eps value, these points are grouped within one cluster. However, the points in the right-side cluster are sparser and the eps value should be increased to 76 to ensure these points are all included in the cluster. Further, if we increase the eps value to 90, the two distinct clusters will be merged as one cluster. At the same time, people's travel trajectory densities captured through social media over different places differ based on the size of a place and the nature of their activities in the

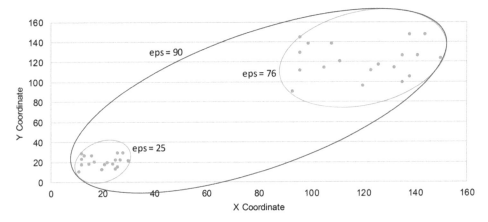

Figure 1. Randomly generated two clusters with different densities.

place. For example, people may have a relatively compact digital footprints at their work place such as their office, and dispersed footprints in a large shopping mall. What is more, the two locations (office building and shopping mall) could be very close to each other as the example (Figure 1). Therefore, an ideal clustering algorithm for digital footprints should not only be able to detect, but also separate different clusters with varying densities.

To mitigate the uncertainty of the clustering results produced by DBSCAN for digital footprints of human movements, this paper develops an improved density-based algorithm based on DBSCAN, named as Multi-scaled DBSCAN (M-DBSCAN) to detected activity clusters at different scales of density and cluster size. Given the historic geo-tagged tweets of an individual, M-DBSCAN can automatically produce a set of clusters and detect appropriate local *eps* and *minpts* values, instead of using one global parameter setting (i.e. the same *eps* and *minpts* value) for each cluster indicating the activity location (zone) the individual visits frequently. The experimental studies demonstrate that the proposed M-DBSCAN algorithm outperforms the classic DBSCAN method and its improved variation, varying DBSCAN (VDBSCA; Liu *et al.* 2007), for activity zone detection from individual geo-tagged tweets with varying-density distributions. Especially, while DBSCAN with inappropriate *eps* and *minpt* values may treat some points (i.e. tweets) in the same activity zone as noise (e.g. tweets posted at the parking lot or pool area of an apartment; Figure 4), our algorithm can significantly reduce the noise ratio (the proportion of tweets not included in any cluster) by identifying all points capturing the activities performed in each zone. Finally, using the proposed M-DBSCAN algorithm on different geo-tagged tweets produced by a large number of users in Washington, DC (DC), this paper explores the range and distribution patterns of appropriate parameter (*eps* and *minpts*) settings for activity zone detection of individuals.

In the next section, we introduce relevant work of using social media data for human mobility studies and state-of-the-art work on DBSCAN for spatial data clustering. Section 3 discusses the uncertainty during the process of regular activity zone detection. Section 4 introduces the datasets and Section 5 presents the M-DBSCAN algorithm and demonstrates how the proposed algorithm can effectively detect activity zones of varying densities. Section 5 compares the results of M-DBSCAN and DBSCAN to develop a general guideline to select the values of DBSCAN parameters for highly varying densities of digital footprints. In the conclusion, we summarize and provide insights about the effectiveness of the M-DBSCAN, and a recommendation of selecting *eps* value for DBSCAN.

2. Related work

Social media recently emerges as a new data source for examining human movement behaviors and mobility patterns. This new trend is driven by the fact that social media data have several unique advantages (Huang and Wong 2016, p. 1) a large number of study subjects. For example, Twitter data have recorded surprising snapshots of human daily movements at a large scale (e.g. 500 million registered users publishing 400 million tweets per day at the year of 2013; Morstatter *et al.* 2013, p. 2) publicly accessible. Using the Twitter application programming interfaces (APIs), we are allowed to retrieve up to 1% of Twitter data (Morstatter *et al.* 2013); and 3) long-term trajectories are available. These data are generated continuously, reflecting the on-going societal situations. They cover extensive temporal scales and provide near real-time information.

As a result, social media is largely used in various human mobility studies (Huang and Wong 2015, Steiger *et al.* 2015, Luo *et al.* 2016, Shaw *et al.* 2016). However, the nature of social media imbues its data with certain blind spots. For example, such data rarely offer much information on the background (e.g. socioeconomic status, demographic informa-tion, or home location) of users. This triggers another research line focusing on social media data mining that intends to infer the background of social media users. In those studies, an important step is detecting the regular activity regions using spatial cluster-ing method that a social media user frequently visits. Among various spatial clustering algorithms, DBSCAN (Ester *et al.* 1996) is often used to detect clusters from digital footprints. However, while we do not need to supply the number of clusters before clustering begins like the K-means method, DBSCAN still requires *minpts* and *eps* values to be provided. While previous studies use a value of 4 and 20 for *minpts* and *eps* respectively (Ester *et al.* 1996, Zhou *et al.* 2004, Hu *et al.* 2015), whether they are optimal for sparse digital footprints are unknown.

In addition, onefold DBSCAN fails to identify clusters from datasets of varying density. To overcome such a limitation, Liu *et al.* (2007) introduced a new algorithm named VDBSCAN to detect clusters with varied densities using the following steps:

(1) Given a *k* value, compute the distance of each point to its kth nearest neighbor (k-dist), and generate a k-dist plot for all points based on their k-dist value sorted ascendingly (Figure 3).
(2) Determine the number of densities intuitively based on the k-dist plot.
(3) Select parameter *eps* for each density.
(4) Scan all the points and cluster points of varied densities with corresponding *eps* using DBSCAN.
(5) Finally, display the valid clusters corresponding with varied densities.

However, the value of parameter *k* has to be manually supplied and no logical procedure or principle is used for determining its appropriate value. As a result, Chowdhury *et al.* (2010) developed a new method to detect the value of parameter *k* automatically based on the characteristics of the datasets. Specifically, given a set of points (P_i), six steps are used to calculate the parameter *k*:

(1) For each point P_i, calculate average distance $d(P_i)$ from P_i to all other points.
(2) Compute the average of all $d(P_i)$ as avg(d).
(3) For each P_i in the datasets, draw a circle with P_i as the center and avg(d) as the radius.
(4) For every circle, find the point nearest to the circumference of each circle, and tag this point as T_i.
(5) Identify the position of T_i as $T_i(Pos)$ relative to the P_i for that particular circle.
(6) Determine the mode of $T_i(Pos)$ and use this value as the value of parameter *k* in the k-dist plot.

While the method works well on a small number of points that are relatively close with each other (e.g. Figure 1), our experiment results indicate that it cannot identify a reasonable *k* value for digital footprints. On one hand, such data could be geographically distributed in a large area, resulting in a large average distance $d(P_i)$ and overall average distance avg(d) at the first and second steps, respectively. As a result, a very large *k* would be detected. On the other hand, trajectory data could be generated at several places with each place including a set of points very close or even overlapping with each other, and accordingly a very small *k* value may be identified. However, the number of points (cluster size) in each activity zone are highly different. In fact, it is observed that most of the users have a primary cluster including a large number of the digital footprint points (46.4% of clustered points on average), and several relatively smaller clusters (Section 6.1). While a large *k* is expected for detecting the primary cluster, relatively small *k* values should be used for the remaining clusters. Therefore, we argue that different local *k* values instead of a global value should be used for detecting clusters with points of varying densities and multiple scales, and heterogeneous spatial distribution. In addition, many existing methods (Liu *et al.* 2007, Chowdhury *et al.* 2010) manually identify *eps* value from the k-dist plot, which decreases *eps* accuracy and clustering efficiency (Parvez 2012).

A few multi-level or hierarchical methods have been proposed to produce clusters with varying densities. To represent the density-based hierarchical clustering structure of a data set, OPTICS (Ankerst *et al.* 1999) constructs a plot of reachability distance, which in turn can be used to extract clusters of different density accordingly. However, OPTICS does not explicitly produce clusters. Wang *et al.* (2016) generated two different *minpts* values by analyzing an adjacency list graph and found more meaningful clusters of varied densities on the sample data set. However, this algorithm still needs to supply *eps* value. Karami and Johansson (2014) combined Binary Differential Evolution and DBSCAN algorithm to automatically tune the best combinations of *eps* and *minpts* for varying data distributions. Campello *et al.* (2013) pointed out the interrelationship between outliers and clusters, and proposed a hierarchical density estimation method for outlier score calculation, which unified the detection process of both local and global outliers. The hierarchical clustering method first computes a hierarchy of mutual reachability distances to represent the distance between a pair of objects, and detects connected components and noise objects at each hierarchical level. Next, an improved hierarchical clustering method was further proposed to simplify the representation of clustering hierarchy for detecting the most significant clusters with different densities by maximizing their overall stability (Campello *et al.* 2015). These three studies were evaluated and achieved good classification results with general datasets. However, these algorithms are very complicated to generate density estimates and structure (e.g. hierarchy), and their performance and applicability on large-scale and complex datasets is unknown.

To address the limitations of existing work of using a density-based approach for digital footprints, and fully automate the clustering process, we develop an M-DBSCAN algorithm (Algorithm 1 and Algorithm 2 in Section 5). It can aggregate spatial points derived from geo-tagged social media data, into clusters of varying spatial distribution and densities with minimal user inputs.

3. Uncertainty of detecting activity zones using digital footprints

In this work, we define an activity zone as the activity region or area that an individual frequently appears or visits. In other words, activity zones should reflect actual human activity space and include frequent (or regular) activities. This section discusses the potential sources of uncertainty introduced in the process of detecting regular activities from four aspects, including 1) data sources, 2) methods for identifying activity zones, 3) representation of the activity zones, and 4) inference of activity zone type. While this paper focuses on addressing the uncertainty posed by methods for detecting activity zones, we elaborate each source as below.

3.1. *Uncertainty from data sources*

The digital footprints are sparse in nature and include a relatively small number of points. Footprint points are captured only when users post a message on the social media platform, and also enable the location-tagging service. As such, compared to spatial trajectory data collected through GPS devices or travel diary establishing the movement trajectories with great detail, using social media data needs to handle 'hit or miss' situations (Huang and Wong 2016). For example, 'check-in' data captured from social media (e.g. Foursquare) may not reveal individual's regular movements and trajectories of individuals. This is because many users are less likely to check in at the regular activity locations (e.g. work), but more likely to check in the places for random activities, such as dining and entertainment (Noulas *et al.* 2012b, Liu *et al.* 2015).

Other social media platforms for people sharing personal experiences and thoughts, such as Twitter and Facebook, can automatically include location to each message as a geo-tag once precise location service is enabled. Accordingly, digital footprints recorded by these platforms have potential to capture human daily movements more comprehensively (Liu *et al.* 2015). Still, people use such platforms more likely in certain places (e.g. metro/bus stations) and time periods (e.g. non-working time during lunch or in the evening), contributing to the spatial and temporal biases of the data (Huang and Wong 2016). In other words, digital footprints might not capture daily regular activity zones (e.g. work), and include irregular activities (e.g. dinning at certain places) instead, introducing noise or uncertainty.

Additionally, location position accuracy and uncertainty are always a part of the conversion when leveraging digital footprints for activity pattern studies. To attach additional location context to a message, users can add a general location (e.g. 'Madison, Wisconsin'), a specific business, landmark, or other POI, or precise location of the mobile device while posting the message. A general location (e.g. city's location), or a specific POI near to a user's activity can mask the true location of the message – as a proxy of user's activity location. Additionally, the precise locations of social media messages (e.g. geo-tagged tweets) are primarily acquired through three methods: GPS, cell tower triangulation, and WiFi hotspots. Different devices with varying positioning methods have its own levels of precision. For example, Zandbergen (2009)'s work reveals that the location of an iPhone 3 achieved an average accuracy of 8 m with GPS, 74 m with WiFi and 600 m with cell tower positioning. As such, location accuracy could be varying across messages, and the detected activity zones may not accurately reflect the areas the user frequently visits.

3.2. *Uncertainty from activity zone detection method*

Digital footprints are recorded as a series of ST points that represent the varying locations of an individual has visited. The locations are resulted from both the regular and random visits. To detect the regular activity zones of an individual, spatial clustering analysis is applied to detect the places that an individual's regular activities take place, and identify outliers (noise) associated with irregular activities. Three major uncertainties could be introduced in this process, whereas the first uncertainty is influenced by the choice of clustering algorithm. Using different clustering algorithms, the number and the shape of clusters are highly varying. Additionally, clustering algorithms usually need to provide one or more input parameters greatly impacting the clustering process and produced results (Moreira *et al.* 2013). Therefore, even using the same clustering algorithm, the results are most likely different depending on the parameterizations of the algorithm (e.g. different *eps* and *minpts* values for the DBSCAN; Figure 4).

The second uncertainty results from the determination of the membership of footprint points while grouping them into different clusters, which are far less than straightforward and objective. For example, points A-F (Figure 4(f)) would be normally and ideally classified as outliers since they are relatively far away from other points in the cluster. However, based on the geographic environment interpreted from the Google Earth map, these activities captured through the online platform are performed in neighborhood of an apartment: points B-D are located at the pool associated with the apartment community, F is recorded at the parking lot, and finally A and E probably are on the entry to the apartment. Therefore, we can reasonably argue that these points are part of the activity zone of the individual, and a spatial clustering algorithm should capture these points instead of treating them as noise. Especially, many users only have a small amount of ST points recorded, the portion of noisy points should be kept minimal to discover meaningful human mobility patterns. On the contrary, while sometimes it is ideal to aggregate 'outlier' (or noise) points into a cluster, we may also need to collapse distinctive clusters capturing different types of activities and assign different clustering memberships to each point. For example, activities performed within two spatially close buildings but with different functions (e.g. work and eating) should be ideally separated into two zones to examine an individual's mobility patterns. As such, the evaluation of a spatial clustering method is rather complex, and it should depend on how we define the noise and how we define the membership of a point in clusters among the trajectory points.

The third uncertainty is introduced during the process of discriminating a regular activity cluster from a random one among the clusters with a wide range of point sizes detected in each cluster. A common approach defines the minimal number of points (*mpts*) as a threshold to identify regular activity places and remove random activity places (Huang and Wong 2016). Specially, a cluster with the number of points < *mpts* are treated as noise and thus discarded. It is worth noting that *mpts* is different from *minpts* in the DBSCAN algorithm. While *minpts* is used during the clustering process to differentiate core points, boarder points, and noise of clusters, *mpts* removes random activity zones after clustering. A lower *mpts* value captures more clusters as the regular activity zones, and therefore may provide more detailed information to describe the daily movements of an individual and to perform more advanced human mobility analysis (e.g. daily trajectory analysis). However, it could include noisy clusters indicating random activities. A more complex solution should consider both the number of points and the temporal information of the points in a cluster. While the total point size in a cluster is larger than *mpts*, they could be generated just in one day. Therefore, this cluster should not be considered as a regular activity place as well. As such, we can define another temporal threshold as the minimum number of days (*mday*) to further control the selection of regular activity zones. Similar to the cluster size threshold (*mpts*), the choice of this temporal threshold value impacts the activity zone identification, therefore introducing uncertainty.

3.3. *Uncertainty from representation of the activity zone*

3.3.1. *Activity zone boundary*

The next step of examining the activities with digital footprints typically moves to the delineation of the activity boundary of each place detected from spatial clustering as an activity zone, which identifies the potential region and measures the size and trend of the area an individual moves at the place. However, the detected regions are influenced by the choice of the boundary reconstruction algorithms, such as convex hull, non-convex (concave) polygon, minimum boundary box, circumscribed circle, and standard deviational ellipse, resulting in highly varying polygon shapes and sizes (Figure 2). For example, while only one convex hull can be detected for a set of points, varying non-convex boundaries can be produced depending on different algorithms and associated input parameterizations. The chi-shape algorithm, a widely used algorithm for generating the concave hull, requires to provide the maximum length of border edges (*l*) as the input parameter (Duckham *et al.* 2008), making the produced boundaries sensitive to the choice of *l* values. Smaller *l* values may generate a boundary better characterizing and generalizing the activity regions where most of the points are concentrated. However, it could underestimate the area of an activity zones which an individual visits comparing with the shapes delineated by other methods (e.g. convex hull; Figure 2).

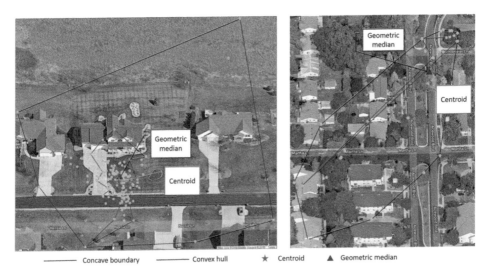

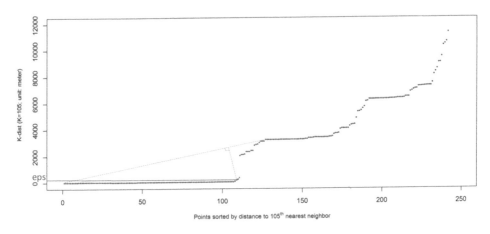

Figure 2. Boundaries and representative locations of two clusters using digital footprints.

Figure 3. Generation of *eps* value based on k-dist plot.

3.3.2. *Representative location*

To construct a two-dimensional (2D) or three-dimensional (3D) space-time (s-t) path displaying movement trajectory, a representative location should be selected for each activity zone. By connecting consequent locations representing the activities occurred in different time period, s-t paths can be established. As such, the derived s-t paths depend on the selection of representative locations, and inappropriate representative locations could result in inaccurate depiction of regular daily movements. While the centroid (blue stars in Figure 2) of all points within a cluster is commonly used to as the representative point of the cluster (Huang and Wong 2015), it is not real trajectory point generated by an individual through online platforms. In some cases, centroids could be very close to a real point (green points in Figure 2(a)). However, they could be located far away from any real trajectory point (Figure 2(b)), and therefore resulting s-t paths may be less precise while

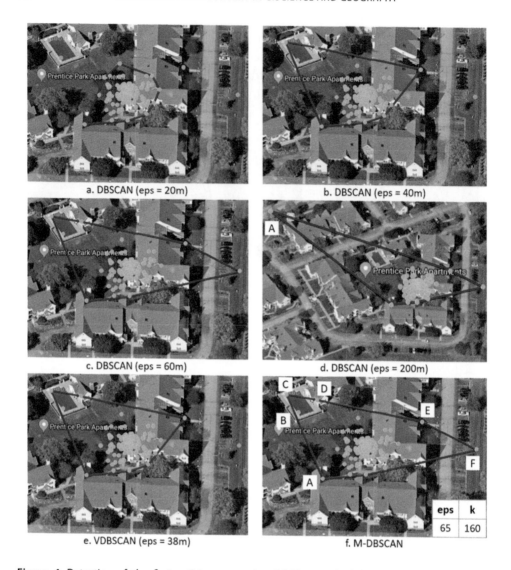

Figure 4. Detection of the first activity zone using DBSCAN with different *eps* values, VDBSCAN and M-DBSCAN for one selected Twitter user.

representing the individual's daily movement trajectory. To cope with this issue, geometric median, the point with minimal total distance to other points in a cluster, is used to represent the cluster (red triangles in Figure 2; Jurgens 2013). Geometric medians ensure that the selected representative locations are meaningful to the individual. Nevertheless, they still cannot guarantee that its location represents the exact spot where an individual stays most in an activity zone (e.g. the room where the individual lives in an apartment, works in an office or has lunch in a restaurant), resulting uncertainty while representing the movement trajectory of the individual using these locations.

3.4. *Activity zone type inference*

Through spatial clustering, a set of clusters are derived from an individual's daily footprints recorded by social media platforms with each cluster covering a region (i.e. activity zone) which this individual regularly visits. To identify the activity zone types, many scholars rely on land-use data as functionality of the activity zone (i.e. land-use type) which is correlated with people's activity there (Huang *et al.* 2014, Huang and Wong 2016). For example, people usually live in residential area and work in industrial or commercial area. Besides, the spatial distributions of POIs within or near the activity zone can indicate different types of people's activities or urban functional zones (Yuan *et al.* 2012, Jiang *et al.* 2015, Gao *et al.* 2017, Cai *et al.* 2018). For example, the inclusion of several restaurants within the activity zone might indicate an eating activity there.

However, the inference of the actual activities performed in activity zone is quite challenging and the accuracy of the inference based on land use and/or POIs cannot be guaranteed. For example, a building in a mix urban area may be used as both residential and commercial usage with upstairs associated with home activities, and downstairs primarily used for eating, shopping, entertainment or other activities. Given an activity zone (e.g. shopping mall), many different types of POIs could be returned. However, normally the primary POI type with the largest count (e.g. shopping store) is used to infer the activity type for an individual' activities on this zone, while other potential types (e.g. restaurant) are often ignored. As such, the results of activity zone type inference are highly uncertain and depend on the inference methods and underlying landscape and surrounding environment where the zones are located.

4. Datasets

This study uses Twitter data, publicly accessible digital footprints, to evaluate M-DBSCAN, and examine the sensitivity of the *eps* and *minpts* values to identify an individual's regular activity zones. Specifically, users in Madison, Wisconsin (Madison), and Washington DC (DC) were chosen, and Twitter's streaming API was used to archive geo-tagged tweets posted within Madison, and DC for a 3-month period. 13,922 unique users were identified within the boundary box of Madison. For DC, 14,066 unique users were identified with each user including 'Washington, DC' as their location information while signing up the Twitter account. For these users, we then re-harvested their historic tweets with the Twitter's search API, which allows to retrieve a maximum of 3,200 most recent tweets for each user.

We used the datasets generated by Madison users to demonstrate and evaluate M-DBSCAN. To make sure each user has sufficient trajectories for clustering discovery, we only selected the users who posted more than 200 geo-tagged tweets, resulting in only 49 eligible users. Since DC has a large number of social media users, we select DC users to examine the activity patterns based on digital footprints. We discard users with geo-tagged tweets less than 50 as these users may have inadequate trajectory points to unfold their activity patterns. Additionally, some users may have highly spare and diversified trajectories, and therefore no spatial cluster can be discovered from their data based on the DBSCAN algorithm. These users were removed from our future analysis. Finally, 3,351 DC users are used for our designed experiments in Section 6.

5. M-DBSCAN

In the following, we introduce the M-DBSCAN algorithm, and then validate how the proposed algorithm can effectively detect activity zones of varying densities subjectively and objectively.

5.1. *M-DBSCAN algorithm*

Given a set of digital footprint points, M-DBSCAN includes three steps to iteratively detect different local k and *eps* values for different clusters:

Step 1. Calculate the optimal k value with Algorithm 1.
Step 2. Calculate an eps value by partitioning and analyzing the k-dist plot based on the optimal k value (Figure 3).
Step 3. Generate clusters by executing DBSCAN using the k value generated in Step 1 as *minpts* and the *eps* value in Step 2 as *eps*.

Above three steps will be repeated until no effective k can be found in Step 1 to ensure that all the clusters are detected properly using corresponding local *eps* values. Algorithm 1 describes the detection of k value (Step 1; Algorithm 2, line 5). Similar as Chowdhury *et al.* (2010), k is determined based on the average mutual distance d(P_i) of every two points to be clustered (line 1). In order to make sure that we compute a local k for points in each individual activity zone, rather than deriving a large global k for all points, we only calculate the average distance of each point to its near neighbors, instead of all other points. Specifically, to calculate d(P_i) for each point (P_i), any point not within a radius of (R) of P_i is considered as irrelevant and discarded. The larger R, the larger d(Pi) is calculated for each point (Pi), which will produce a larger avg(d) and thus a larger k. Similarly, the smaller R, the smaller k will be calculated. In other words, R's value influences the detection of k value and clusters accordingly.

While there is no existing guideline to select an appropriate R value, previous study shows that distance from the typical American's house to the edge of his community is between 520 and 1060 m (Donaldson 2013). In other words, activities performed within one zone most likely are within a radius of 1060 m. Therefore, we use 1060 m as the value of R in our study and two points more than 1060 m away are considered as two different types of activities. Therefore, mutual distances larger than 1060 m are discarded when calculating d(P_i) and avg(d).

Next, for each P_i, its neighborhood is obtained by collecting all the points within a radius of avg(d) centered on P_i. The amount of points within the neighborhood of P_i is recorded as k_i. A list [k_i] for all points are collected, which can be used to calculate the frequency (or count) of each k_i value. These unique k_i values are recorded as [k_j] and their frequencies are [f_j] (line 4). Then, we delete k_j with small f_j value by selecting the top five k_j with the largest f_j values (line 5). Next, [k_j] are sorted based on their frequency to find the largest k_j, which is selected as k value for the set of points (line 7). Based on our experiments (Section 5.2), each user has activity clusters of varying size (the number of points in a cluster) and density. This step iteratively identifies the optimal k to detect the current largest cluster(s).

Algorithm 1: Get k and corresponding *eps*

Input: The amounts of remaining points (RP) and rank

Output: k and corresponding *eps* based on k-dist plot

1: calculate avg(d) by averaging the average mutual distance of P_i to all other points within a radius R

2: calculate the amounts of points [k_i] within avg(d) of P_i

3: while (t < rank)

4:      generate map [(k_j, f_j)] (f_j is the frequency of k_j value)

5:      sort [(k_j, f_j)] descending by f_j

6:      maintain the anterior of [(k_j, f_j)] where f_j-f_1 < 5

7:      sort [(k_j, f_j)] descending by k_j

8:      delete (k_1, f_1) from [(k_i, f_i)]

9:      t ++

10: end

11: if k cannot be found

12:      return −1

13: sort distances ([d_i]) between P_i and its kth nearest neighbor (k-dist plot)

14: find start points (S_1, S_j) of the first two flat parts on k-dist plot

15: find the knee point S_{knee} with the maximum perpendicular distance to the connecting line of S_1 and S_j

16: return the *k* value and *eps* value, which is the distance from the S_{knee} to its kth nearest neighbor

Step 2 calculates corresponding *eps* value based on the local optimal *k* value. Instead of manually identifying the sharp change of k-dist plot for *eps* (Liu *et al.* 2007, Elbatta and Ashour 2013), we propose a novel method to automatically detect its value (Algorithm 1, lines 13–15). The distances from each P_i to its kth nearest neighbor are sorted ascendingly (an example with k = 105 shown in Figure 3). The first two parts of the point array with relatively flat ascending trend are then detected. The start points of the two parts are connected to find the farthest point between the two start points, known as S_{knee}, which has the maximum perpendicular distance to the connecting line (Line 15). S_{knee} indicates the sharpest change of plotted distances along the connecting line, and thus is selected to derive the *eps* value (Line 16).

Algorithm 2: M-DBSCAN

Input: a set of points representing geo-tagged tweets

Output: representative clusters signified as lists of tweets (RC)

1: while the amounts of remaining points (RP) are no less than 4

2:      if RC remain the same

3:            rank = 1

4:      else   rank ++

5:      get K and corresponding eps (Algorithm 2)

6:     if eps < 0

7:          break

8:     if eps <20

9:          continue

10:    apply DBSCAN (minpts = K, eps = eps)

11:    add newly generated representative clusters to RC

12:    delete newly clustered tweet points from RP

13: end

Given a set of points representing an individual's daily trajectories, Algorithm 2 describes the entire process of M-DBSCAN for detecting his or her activity clusters. A loop is used to iteratively obtain the optimal *eps* value representing the radius of the current most prominent cluster, and *k* representing the minimum number of the points in the cluster (Line 5). They are taken as the values of *eps* and *minpts* respectively to conduct DBSCAN on the point set (Line 10). Then clustered points are deleted from the point list which will be further clustered during the next round (Line 12). The process is repeated until there are less than four points left since a cluster with three or less points are not representative enough to be considered as an activity zone, or no more cluster can be found using each possible *k*.

5.2. Demonstration of the M-DBSCAN using twitter data

5.2.1. Detection of activity zones of varying densities

A Twitter user is selected to demonstrate and evaluate the effectiveness of the M-DBSCAN on automatically detecting activity zones with varying *eps* and *minpts* values. Figure 4 displays the results of the first activity zone detection using DBSCAN with different *eps* values and M-DBSCAN for the selected Twitter user. Based on the surrounding environmental information indicated from the Google Earth map, we can infer that the activity zone is located within an apartment community and ideally footprint points capturing the activities within this region should be all considered as one cluster. With a small *eps* value (e.g. 20 and 40 m), DBSCAN can detect most of the points distributed in the apartment area but not those points in the pool, lawn and parking area. By using the *eps* value of 60 m, all the activities are captured in the cluster. However, further increasing the *eps* value to 200 m, an outlier (point A in Figure 4(d)) located at the street far away from the apartment building is included in the cluster. Similarly, VDBSCAN detects an *eps* value as 38 m, which cannot capture all the activities as well (Figure 4(e)). M-DBSCAN automatically detects the *eps* value as 65 m and obtains the ideal clustering results (Figure 4(f)).

The second example of activity regions (Figure 5) is also a residential community but with single house families. Using DBSCAN with *eps* value equal to 40 m, four small clusters (A, B, C and D) are detected (Figure 5(a)). Increasing the *eps* value, small clusters are merged as a big one (cluster A) and an additional small cluster (cluster B) across the transmeridional main avenue is detected (Figure 5(b)). Next, with 200 m as the *eps* value (Figure 5(c)), all the activity points are captured in two detected clusters. Finally, further increasing the *eps* value to 320 m merges the two clusters (Figure 5(d)), which ideally should be separated as there is a wide avenue between them. As the DBSCAN with an *eps* value of 40 m, VDBSCAN detects

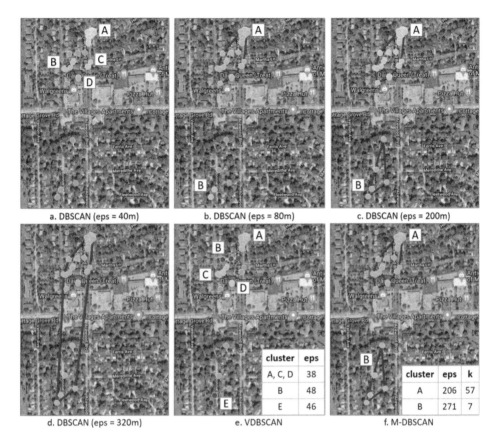

cluster	eps
A, C, D	38
B	48
E	46

cluster	eps	k
A	206	57
B	271	7

a. DBSCAN (eps = 40m)   b. DBSCAN (eps = 80m)   c. DBSCAN (eps = 200m)
d. DBSCAN (eps = 320m)   e. VDBSCAN   f. M-DBSCAN

Figure 5. Detection of the second activity zone using DBSCAN with different eps values, VDBSCAN and M-DBSCAN for the selected Twitter user.

fragmentary activity zones (Figure 5(e)) using multiple *eps* values from 38 to 48 m, within the northern area. For the southern area, VDBSCAN is able to detect an additional activity zone (cluster E). Contrastively, M-DBSCAN identifies two optimal clusters using a local eps values of 206 and 271 m, respectively. The detected clusters are similar as DBSCAN using an *eps* value of 200 m (Figure 5(f)).

5.2.2. Detection of activity zones of low densities

Besides enabling inclusion of more eligible footprint points, M-DBSCAN can detect clusters of largely diverse densities (Figure 6), some of which are missed by either DBSCAN or VDBSCAN. When using 60 m as *eps*, the residential area (blue boundary in the inset map of Figure 6) demonstrated in Figure 4, can be detected while a prominent shopping area is missing (Figure 6(a)). When using 500 m as eps, the shopping area can be detected (cyan boundary) while the residential area is much enlarged that two noise points are also included (Figure 6(b)). VDBSCAN cannot detect the shopping area either (Figure 6(c)). However, M-DBSCAN can detect both the residential area and the shopping area without evident noise (Figure 6(d)).

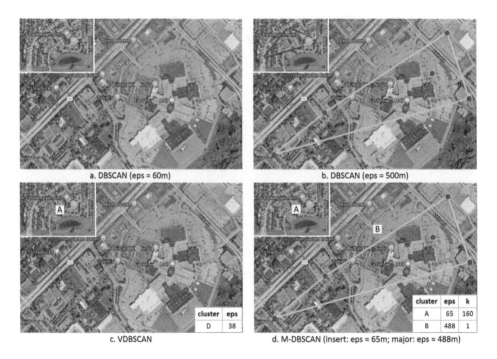

Figure 6. Detecting activity zones of extremely diverse footprint densities using DBSCAN with different eps values, VDBSCAN and M-DBSCAN for one selected Twitter user; insert maps at the upper-left show the detection of the residential activity zones for the selected user as Figure 4; main maps show the activities around the shopping mall area.

The three examples demonstrate that different activity zones include highly diversified activities. While some activities may occur in a compact space (e.g. apartment) which can be detected with smaller *eps* values (Figure 5), an activity zone (e.g. shopping) may include scattered activities in a large space (e.g. mall), and therefore a large *eps* value should be used to capture all of them (Figure 6). The developed M-DBSCAN can automatically detect varying *eps* values for digital footprints of varying density and capture the maximal scope of activities.

5.3. Clustering evaluation with rand index and adjusted rand index

Rand index (RI; Rand 1971) is an evaluation measure for a clustering problem in terms of agreement or disagreement between object pairs in two partitions (Walde 2006). In this measurement, if a pair of objects are assigned to the same class or they are assigned to different classes using two different partition methods, the assignment is considered as an agreement (A), otherwise it is considered as a disagreement (D). The similarity of two partitions can thus be evaluated by measuring the overlap of A versus D.

Given a set of objects $O = \{O_1, O_2, \ldots, O_n\}$, the $RI(C, M)$ between a clustering result $(C = \{C_1, C_2, \ldots, C_x\})$ and the manual classification $(M = \{M_1, M_2, \ldots, M_y\})$ is defined as:

$$RI(C,M) = \frac{\Sigma_{i<j}^{n} \gamma(o_i, o_j)}{\binom{n}{2}},$$

where n is the number of objects, $\binom{n}{2}$ is computed as n(n-1)/2, and

$$\gamma(o_i, o_j) = \begin{cases} 1 \text{ if there exist } C_A \in C, M_B \in M \text{ such that objects } o_i \text{ and } o_j \text{ are in } C_A \text{ and } M_B, \\ 1 \text{ if there exist } C_A \in C, M_B \in M \text{ such that } o_i \text{ is in both } C_A \text{ and } M_B \\ \quad \text{while } o_j \text{ is in neither } C_A \text{ or } M_B, \\ 0 \text{ otherwise.} \end{cases}$$

Based on above definition, $RI(C,M)$ ranges from 0 to 1. A larger $RI(C,M)$ indicates that partition C is in higher agreement with partition M, indicating the clustering result is more similar to the ground truth (manual classification) and thus is better. If C perfectly agrees with M, $RI(C,M)$ achieves 1. However, the problem with the RI is that its expected value can vary when C and M are both partitioned randomly, which is expected to be constant when both partitions are generated with a random classification model. To solve this problem, the adjusted rand index (ARI; Hubert and Arabie 1985) was therefore developed by introducing the expected similarity of all pair-wise comparisons between C and M partitioned by a random model to the calculation. While the ARI has the maximum value 1, it produces a value of zero for all random partitions. Similarly, a larger ARI means that a clustering result has more agreement with the manual partitions (i.e. ground truth). Hereby, an ARI between C and M ($Rand_{adj}(C,M)$) is defined as (Hubert and Arabie 1985):

$$Rand_{adj}(C,M) = \frac{\Sigma_{i,j}\binom{t_{ij}}{2} - \frac{\Sigma_i\binom{t_i}{2}\Sigma_j\binom{t_j}{2}}{\binom{n}{2}}}{\frac{1}{2}\left(\Sigma_i\binom{t_i}{2} + \Sigma_j\binom{t_j}{2}\right) - \frac{\Sigma_i\binom{t_i}{2}\Sigma_j\binom{t_j}{2}}{\binom{n}{2}}},$$

Where t_{ij} denotes the number of objects in common between $C_i(\in C)$, and $M_j (\in M)$, t_i and t_j indicates the number of objects in common between C_i and M, and between M_j and C respectively.

ARI is widely used to validate clustering results by measuring the agreement between two partitions: one is generated by the clustering algorithm, and the other is produced by external criteria (e.g. manual classification), especially when the two partitions contain different numbers of clusters (Yeung and Ruzzo 2001, Walde 2006, Santos and Embrechts 2009). We thus utilize it to evaluate clustering results of different spatial clustering methods and to validate the effectiveness of the proposed M-DBSCAN model.

Firstly, we sort Twitter users in Madison based on the number of distinct dates when they posted online messages, and select the top 10 users to manually classify their geo-tagged tweets (i.e. partition M). Based on the uncertainties of activity zone detection discussed in Section 3, we follow three principles while manually classifying the geo-tagged tweets (Figure 7) as an activity zone: 1) tweets intensively (≥4 points; i.e. frequent tweets) located at a specific place (e.g. mall, restaurant, school, apartment)

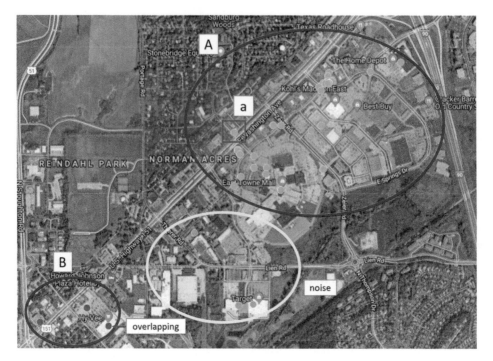

Figure 7. Demonstration of manual classification criterions.

or neighboring places for the same type of activities (e.g. shopping, eating) are considered as one cluster (i.e. cluster A and B; Note while it looks like only three points are displayed in the cluster B, there are repeated points overlapping at the same location as labeled in Figure 7); 2) tweets located outside but near a cluster, which represent the same activity as these tweets within the cluster (e.g. tweets located at the parking lot around a shopping mall; i.e. point a in Figure 7), should also be included in the cluster; 3) infrequent tweets (<4 points) located at a specific place or spread across multiple neighboring places for the same activity type (i.e. noise points in Figure 7) and frequent tweets located at distinct places for different activities (e.g. two tweets located at a restaurant and another two at a post office) are not clustered.

Next, we calculate $RI(C, M)$ and $ARI(C, M)$ for each selected user and the classification results (partition C) generated by different clustering methods including M-DBSCAN, VDBSCAN, and DBSCAN using different eps values. The averaged values of RI and ARI (i.e. $\overline{RI}(C, M)$ and $\overline{ARI}(C, M)$) for the 10 users are then derived (Table 1). During the calculation, all noise points are considered as one cluster and assigned to the same cluster

Table 1. Rand index and adjusted Rand index evaluation results using different clustering methods.

Cluster method	M-DBSCAN	VDBSCAN	DBSCAN				
eps (m)	N/A	N/A	40	80	100	200	300
$\overline{RI}(C, M)$	0.966	0.927	0.930	0.953	0.945	0.951	0.949
$\overline{ARI}(C, M)$	0.964	0.915	0.858	0.940	0.905	0.928	0.923

number (i.e. 0) as the classification consistency of each pair of points should be considered to measure the agreement between C and M.

The evaluation results show that M-DBSCAN is able to detect the activity zone clusters with the highest agreement with the manually identified activity zones (i.e. ground truth). Specifically, the evaluation results show that M-DBSCAN generates a higher $\overline{RI}$ (C, M) value than any other clustering method, which indicates its effectiveness of separating regular and random activity points. However, the differences of $\overline{RI}$ (C, M) between M-DBSCAN and other methods, especially DBSCAN with eps ≥80m, are rather small. After the adjustment by a random model, a $\overline{ARI}$ (C, M) is generated for each clustering method. These $\overline{ARI}$ (C, M) values are smaller than their $\overline{Rand}(C, M)$ in general. While there is only small decrease over the $\overline{RI}$ (C, M) for M-DBSCAN, the $\overline{ARI}$ (C, M) value decreases more for other methods, such as DBSCAN with eps equal to 40 m and over 100 m. This is because these methods detect a highly varying number of clusters (i.e. either much larger or smaller) as the number of clusters manually labeled.

6. Explore the uncertainty with M-DBSCAN

6.1. Analysis of activity zones

Given a set of 2D points capturing a user's historic trajectories, the proposed M-DBSCAN can automatically detect *minpts* and *eps* for grouping all 2D points into different clusters. Using the data of 3,351 selected DC users, 42,962 clusters are detected in total with each user having one or more clusters (Table 2). A majority of the users (75%) have a cluster number less than 16 and half of the users have a cluster number less than 8 with each cluster having a minimum number of 4 points per cluster (Figure 8). While examining these users with a large number of clusters detected, many users are found out to be also users of other social network platforms (such as Foursquare for checking-in places, and Instagram for sharing photos and videos). These users link Twitter account with other different platforms and the messages posted from these platforms can be automatically published on the Twitter. In fact, 2,192 out of 3,351 users are Instagram users, 977 users are Foursquare users, and finally 759 users are both Instagram and Foursquare users. Through these two platforms, users may generate repeated points of the same location (e.g. check-in the same place repeatedly through Foursquare), resulting a cluster to be formed at the locations. Besides, users of these platforms more likely check in at random places (e.g. restaurants) not visiting frequently in a daily basis, therefore, much diversified movement patterns can be observed.

Table 2. Quantile statistics of the number of activity clusters with different *mpts* value.

mpts value	Percentage of users			
	25%	50%	75%	100%
≥4	4	8	16	133
> 4	3	6	13	88
≥6	3	5	9	63
≥8	2	4	7	52
≥10	2	3	6	41
≥12	2	3	5	30

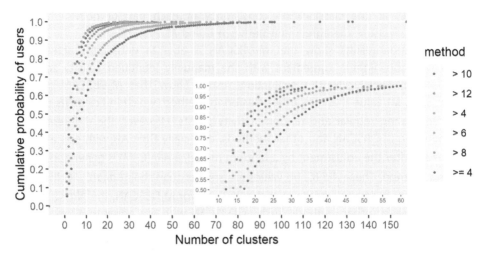

Figure 8. Density of activity zone (cluster) numbers of all users with different value of the minimum number of points (*mpts*) per cluster as the threshold to remove random activity clusters.

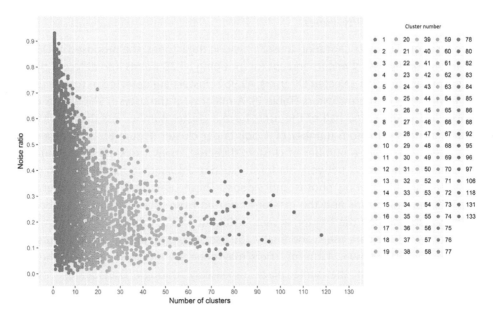

Figure 9. The relationship between the noise ratio and the cluster number of all users with *mpts* = 4.

Figure 9 shows the relationship between the noise ratio and the cluster number generated by all users. Given a set of trajectory points of a user, the noise ratio is the proportion of points not within any cluster, and therefore are considered as noise. For the users having a small number of activity zones (e.g. less than 5) detected, the distribution of noise ratio is relatively random ranging from 0 to approximate 1. With an increasing of activity zone number detected, the noise ratio drops gradually. Especially, for the users with a number of activity zones larger than 40 detected, the noise ratio has a clear distribution pattern with a ratio value mostly less than 40%.

During our experiment, we notice that most of the users have one activity zone, known as the primary activity zone, including the largest number of points (cluster size) and many clusters with relatively smaller point numbers, which most likely capture random activities. This primary cluster captures 46.4% of points that are clustered on average, and the percentage increases to 63.7% combining with the secondary cluster (with the cluster size ranked as second). In the proposed M-DBSCAN algorithm, the minimum number of points (*mpts*) should be included in a cluster is 4. In fact, if we remove the clusters with the number of points as 4 for each user, most of the users (75%) only have less than 13 clusters (second row of Table 2). In other words, each user will have three small clusters with only 4 points on average. For users with a large number of trajectory points, those clusters most likely capture some random activities for users. However, if a user only has a small number of points, these clusters could also record regular activities. Nevertheless, we can use the value of *mpts* as a threshold to remove the potential noisy clusters. If using 12 as *mpts* value, most of the users (75%) have less than 5 clusters detected. This is reasonable since people's regular activity is performed in five regular zones, including home, work, entertainment, eating, and shopping. Therefore, the choice of *mpts* affects the uncertainty of detecting the number of regular activity zones.

Quite interestingly, the primary activity zones are typically detected with a relatively small *eps* value. With an *eps* value larger than 439 m, we capture dispersed clusters with low density of points (Figure 10 left). Most of the clusters (80%) are detected with an *eps* value smaller than 439 m (Figure 10 left). The value of *minpts* detected by M-DBSCAN has a strong positive linear relationship (Pearson's correlation coefficient 0.947) with the size of the cluster to be detected (Figure 10 right). This makes sense as points in a larger cluster typically are denser, which means that each point has more neighbors reachable, and therefore a larger value of *minpts* should be used.

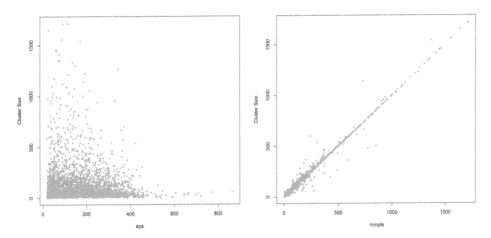

Figure 10. The correlation of *eps* and *minpts* with the number of points (cluster size) in each cluster.

6.2. Selection of eps value for DBSCAN

To explore the range and distribution patterns of appropriate parameter (*eps* and *minpts*) values for activity zone detection of individuals using DBSCAN algorithm, we change the *eps* value from 20 m approximating to the uncertainty in GPS readings, to 400 m. Using a value of 20 m, a large number of small clusters are detected (Table 3). The cumulative distribution of cluster numbers using DBSCAN with an *eps* of 20 m is very different from other methods (Figure 11) confirming that with such a small *eps* value, DBSCAN more likely groups the points into smaller clusters, running the risk of separating the activities of the same zone captured by the geo-tagged tweets into different zones. With the increasing of *eps* value from 20 to 40 m, smaller clusters are merged into larger clusters, resulting in the decrease of the number of clusters detected for all users in total. However, if we continue to increase the *eps* value from 40 to 400 m, an increasing number of clusters are detected, indicating that dispersed clusters with low density of points are identified. Herein, M-DBSCAN has the largest number of clusters detected except for using DBSCAN with an *eps* value of 20 m. This confirms that the M-DBSCAN can capture

Table 3. Results of M-DBSCAN and DBSCAN with different *eps* values and a *minpts* value of 4.

Cluster methods	Average number of clusters	Average noise ratio	Total clusters
DBSCAN (*eps* = 20)	9.75	69.0%	66,042
DBSCAN (*eps* = 40)	10.28	54.1%	35,294
DBSCAN (*eps* = 80)	11.12	49.0%	38,394
DBSCANE (*eps* = 100)	11.32	47.2%	39,187
DBSCANE (*eps* = 200)	11.97	43.1%	41,550
DBSCANE (*eps* = 300)	12.11	40.6%	42,076
DBSCANE (*eps* = 400)	7.82	38.7%	42,190
M-DBSCAN	12.50	32.7%	42,962

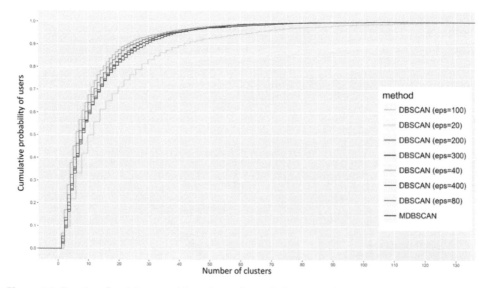

Figure 11. Density of activity zone (cluster) numbers of all users with M-DBSCAN and DBSCAN with different *eps* values and a *minpts* value of 4.

dispersed clusters with low density of points by automatically using larger *eps* values to detect these clusters (Figure 10 left).

With an *eps* value of 20 m used for DBSCAN, a major of points (69%) are detected as noise and therefore are not appropriate for mobility analysis for users on average (Table 3). While increasing *eps* value could reduce the noise ratio on average, the number of discarded points (noise) is still very high. With the increasing of *eps* values from 20 to 300 m, more clusters are detected, which is reasonable as a large *eps* value can help capture clusters of low density. However, if we continue to increase *eps* value from 300 to 400 m, only about eight clusters are detected on average indicating small clusters potentially with different activities being merged into large clusters. In contrast, the proposed M-DBSCAN has the smallest average noise ratio compared with the results detected by DBSCAN with different eps values. With an *eps* value as 300 m, DBSCAN can detect similar number of clusters as M-DSBCAN on average, though with a higher noise ratio.

To further identify appropriate *eps* values for DBSCAN, the two-sample Kolmogorov–Smirnov (KS) statistic (D value in Table 4) is used to test whether two underlying probability distributions of the number of clusters detected for the selected using DBSCAN using varying *eps* values and M-DBSCAN differ. The KS statistics provides a measure of the distance between the empirical distribution function of two samples (e.g. n and m). Given the empirical distribution functions of the first and the second sample $F_{1,\,n}$ and $F_{2,\,m}$, the KS statistic $D_{n,\,m}$ is calculated as:

$$D_{n,m} = sup_x \left| F_{1,\,n}(x) - F_{2,\,m}(x) \right|,$$

where sup_x is the supremum function. The smaller the D value, the better the distribution the two samples fit each other.

The test results in Table 4 show DBSCAN with *eps* values as 200 and 300 m have similar distribution of the number of clusters identified for each user as M-DBSCAN with a small D value. The significance indicated by p-value (0.2953) using 200 m as *eps* is greatly larger than a typical significance level (0.05) and thus we cannot reject the null hypothesis statement that two samples come from the same probability distribution. Besides, the p-value using 300 m as *eps* value is also greater than 0.05, which does not reject the null hypothesis. While increasing the *eps* value from 300 to 400 m, the KS test p-value dropped to 0.00869 and D value increased accordingly. In addition, the D value is larger than the critical value (0.033) at the significance level of 0.05, further confirming the dissimilarity of the distributions of the number of clusters detected by the two

Table 4. Two-sample Kolmogorov–Smirnov test of the probability distribution of the number of clusters using M-DBSCAN and DBSCAN with different *eps* values.

Cluster methods (*eps* in meters)	D value	p-value
DBSCAN (*eps* = 20)	0.148	< 2.2e-16
DBSCAN (*eps* = 40)	0.101	3.775e-15
DBSCAN (*eps* = 80)	0.050	0.0003976
DBSCAN (*eps* = 100)	0.040	0.0102
DBSCAN (*eps* = 200)	0.024	0.2953
DBSCAN (*eps* = 300)	0.032	0.06995
DBSCAN (*eps* = 400)	0.040	0.00869

methods. From the statistic results (Tables 3 and 4), a value between 200 and 300 m is recommended for *eps* while using DBSCAN for detecting activity zones in general.

7. Conclusion and future work

With the advancement of communication and information technologies, social media platforms emerge as new solutions to record people's movements in a daily basis. While exploring mobility patterns, spatial clustering over the digital footprint points is typically used to detect places where an individual regularly visits. Among all spatial clustering algorithms, DBSCAN is widely used as it is effective to detect clusters of arbitrary shape with noise, and only needs to supply *minpts* and *eps* as input, the values of which are relatively easy to determine compared to other algorithms. However, DBSCAN is sensitive to the two input parameters while detecting regular activity clusters from users' social media points, which are sparsely, irregularly distributed in space, and featured of varying densities. DBSCAN and existing improved DBSCAN methods (Ester *et al.* 1996, Ertöz *et al.* 2003, Liu *et al.* 2007, Wang *et al.* 2016) based on a k-dist plot, where k-dist is defined as the distance from each point to its k nearest point, often fall short to identify activity clusters.

This paper develops an improved density-based clustering algorithm based on DBSCAN, named as M-DBSCAN, which can automatically produce a set of activity zones (clusters), and select an appropriate *eps* and *minpts* value for each activity zones. In addition, M-DBSCAN can significantly reduce the noise ratio (the proportion of tweets not included in any cluster) by identifying all points capturing the activities performed in each zone. Using the historical online geo-tagged tweets of users in Madison and in DC, the results of M-DBSCAN and DBSCAN with varying *eps* value indicate that: 1) M-DBSCAN detects more clusters (activity zones) for each user, and results in a lower noise ratio (the proportion of tweets not included in any cluster); 2) A value of 40 m or higher should be used for *eps* in order to reduce the possibility of collapsing the activities of the same zone captured by the social media data points into different zones, and ensure an average noise ratio less than 54% during the clustering process; 3) A value between 200 and 300m is mostly recommended for *eps* while using DBSCAN for detecting activity zones; and finally 4) There is no optimal value for *minpts*. The value of *minpts* detected by M-DBSCAN displays a strong positive linear relationship (Pearson's correlation coefficient 0.947) with the size of the cluster to be detected.

The proposed M-DBSCAN was first evaluated subjectively by analyzing a selected user's activity zones detected by different clustering methods (Section 5.2). Next, we manually identify the daily activity zones of 10 users in Madison as ground truth data to evaluate the effectiveness of M-DBSCAN objectively. While determining the membership of a point (Section 3.2), it is challenging to evaluate whether it should be considered a part of an activity zone or as outlier (Figure 4(c) vs (d)). Similarly, it is quite difficult to identify if one cluster is better to describe an individual's activity zone than to separate it into two clusters or to keep some points as outliers (e.g. Figure 5(d) vs (f)). As such, contextual information derived from the social media contexts (e.g. text messages) and temporal information could be integrated to further improve the detection of activity zones. Finally, this paper explores the uncertainty of activity zone detection using trajectory data of DC users. In the future, footprints of social media users in different cities can be explored to identify activity zones and patterns across regions.

The proposed M-DBSCAN can effectively detect user activity zones based on digital footprints, and has potential to facilitate a wide range of practical applications such as human mobility study, tourism recommender systems, and business site selection, where POIs should be discovered using digital footprints from massive users. Besides, the study results provide general guideline to choose optimal values for *eps* and *minpts* while running the DBSCAN algorithm to detect individual activity zones. In future, we will integrate multi-sourced data (e.g. OpenStreetMap land use, and Google place service data) to detect the activity zone types of each cluster, while allows us to further explore user tweeting behaviors at different types of activity zones and complex mobility patterns, such as trajectory patterns of different socioeconomic and demographic groups.

Disclosure statement

No potential conflict of interest was reported by the authors.

Funding

Support for this research was provided by the University of Wisconsin - Madison Office of the Vice Chancellor for Research and Graduate Education with funding from the Wisconsin Alumni Research Foundation.

ORCID

Xinyi Liu ⓘ http://orcid.org/0000-0002-1431-8816
Qunying Huang ⓘ http://orcid.org/0000-0003-3499-7294
Song Gao ⓘ http://orcid.org/0000-0003-4359-6302

References

Ankerst, M., *et al.*, 1999. OPTICS: ordering points to identify the clustering structure. *In*: ed. *Proceedings of the 1999 ACM SIGMOD international conference on management of data*, 31 May–03 June. Philadelphia, Pennsylvania, USA, 49–60. New York, NY: ACM.

Cai, L., *et al.*, 2018. Integrating spatial and temporal contexts into a factorization model for POI recommendation. *International Journal of Geographical Information Science*, 32 (3), 524–546. doi:10.1080/13658816.2017.1400550

Campello, R.J., *et al.*, 2015. Hierarchical density estimates for data clustering, visualization, and outlier detection. *ACM Transactions on Knowledge Discovery from Data (TKDD)*, 10 (1), 1–51. doi:10.1145/2733381

Campello, R.J., Moulavi, D., and Sander, J., 2013. Density-based clustering based on hierarchical density estimates. *In*: J. Pei, V. S. Tseng, L. Cao., H. Motoda, and G. Xu, eds. *PAdvances in Knowledge Discovery and Data Mining. PAKDD*. Lecture Notes in Computer Science, 7819. Berlin: Springer.

Chowdhury, A.R., Mollah, M.E., and Rahman, M.A., 2010. An efficient method for subjectively choosing parameter 'k'automatically in VDBSCAN (Varied Density Based Spatial Clustering of Applications with Noise) algorithm. *In*: ed. *The 2nd international conference on computer and automation engineering (ICCAE)*, 26–28 February. Singapore, 38–41. Piscataway: IEEE.

Donaldson, K., 2013. *How big is your neighborhood? Using the AHS and GIS to determine the extent of your community* [online]. Available from: https://www.census.gov/programs-surveys/ahs/research/working-papers/how_big_is_your_neighborhood.html [Accessed 30 April 2018].

Duckham, M., *et al.*, 2008. Efficient generation of simple polygons for characterizing the shape of a set of points in the plane. *Pattern Recognition*, 41 (10), 3224–3236. doi:10.1016/j.patcog.2008.03.023

Elbatta, M.T. and Ashour, W.M., 2013. A dynamic method for discovering density varied clusters. *International Journal of Signal Processing, Image Processing, and Pattern Recognition*, 6 (1), 123–134.

Ertöz, L., Steinbach, M., and Kumar, V., 2003. Finding clusters of different sizes, shapes, and densities in noisy, high dimensional data. *In*: D. Barbara and C. Kamath, eds. *Proceedings of the 2003 SIAM international conference on data mining*, 1–3 May. San Francisco, CA, 47–58. Available from: https://epubs.siam.org/doi/book/10.1137/1.9781611972733?mobileUi=0

Ester, M., *et al.*, 1996. A density-based algorithm for discovering clusters in large spatial databases with noise. *In*: E. Simoudis, J. Han, and U. Fayyad, eds. *The second international conference on knowledge discovery and data mining (KDD)*, 2–4 August. Portland, Oregon, 226–231. Palo Alto: AAAI PRESS.

Gao, S., *et al.*, 2014. Detecting origin-destination mobility flows from geotagged Tweets in greater Los Angeles area. *In*: M. Duckham, E. Pebesma, K. Stewart, and A. U. Frank, eds. *The eighth international conference on geographic information science (GIScience'14)*, 24–26 September. Vienna, Austria. New York, NY: Springer.

Gao, S., 2015. Spatio-temporal analytics for exploring human mobility patterns and urban dynamics in the mobile age. *Spatial Cognition & Computation*, 15 (2), 86–114. doi:10.1080/13875868.2014.984300

Gao, S., Janowicz, K., and Couclelis, H., 2017. Extracting urban functional regions from points of interest and human activities on location-based social networks. *Transactions in GIS*, 21 (3), 446–467. doi:10.1111/tgis.2017.21.issue-3

Hu, Y., *et al.*, 2015. Extracting and understanding urban areas of interest using geotagged photos. *Computers, Environment and Urban Systems*, 54, 240–254. doi:10.1016/j.compenvurbsys.2015.09.001

Huang, Q., 2017. Mining online footprints to predict user's next location. *International Journal of Geographical Information Science*, 31 (3), 523–541. doi:10.1080/13658816.2016.1209506

Huang, Q., Cao, G., and Wang, C., 2014. From where do Tweets originate?-A GIS approach for user location inference. *In*: ed. *Proceedings of the 7th ACM SIGSPATIAL international workshop on location-based social networks*, 4 November. Dallas/Fort Worth, Texas. New York, NY: ACM.

Huang, Q. and Wong, D.W., 2015. Modeling and visualizing regular human mobility patterns with uncertainty: an example using Twitter data. *Annals of the Association of American Geographers*, 105 (6), 1179–1197. doi:10.1080/00045608.2015.1081120

Huang, Q. and Wong, D.W., 2016. Activity patterns, socioeconomic status and urban spatial structure: what can social media data tell us? *International Journal of Geographical Information Science*, 30, 1873–1898. doi:10.1080/13658816.2016.1145225

Hubert, L. and Arabie, P., 1985. Comparing partitions. *Journal of Classification*, 2 (1), 193–218. doi:10.1007/BF01908075

Jiang, S., *et al.*, 2015. Mining point-of-interest data from social networks for urban land use classification and disaggregation. *Computers, Environment and Urban Systems*, 53, 36–46. doi:10.1016/j.compenvurbsys.2014.12.001

Jurgens, D., 2013. That's what friends are for: inferring location in online social media platforms based on social relationships. *ICWSM*, 13 (13), 273–282.

Kang, C., *et al.*, 2012. Intra-urban human mobility patterns: an urban morphology perspective. *Physica A: Statistical Mechanics and Its Applications*, 391 (4), 1702–1717. doi:10.1016/j.physa.2011.11.005

Karami, A. and Johansson, R., 2014. Choosing DBSCAN parameters automatically using differential evolution. *International Journal of Computer Applications*, 91 (7), 1–11. doi:10.5120/15890-5059

Kwan, M.-P., 2013. Beyond space (as we knew it): toward temporally integrated geographies of segregation, health, and accessibility: space–time integration in geography and GIScience. *Annals of the Association of American Geographers*, 103 (5), 1078–1086. doi:10.1080/00045608.2013.792177

Lin, J. and Cromley, R.G., 2018. Inferring the home locations of Twitter users based on the spatiotemporal clustering of Twitter data. *Transactions in GIS*, 22 (1), 82–97. doi:10.1111/tgis.2018.22.issue-1

Liu, P., Zhou, D., and Wu, N., 2007. VDBSCAN: varied density based spatial clustering of applications with noise. *In: ed. 2007 international conference on service systems and service management*. Chengdu, China, 1–4. Piscataway: IEEE.

Liu, Y., *et al.*, 2015. Social sensing: A new approach to understanding our socioeconomic environments. *Annals of the Association of American Geographers*, 105 (3), 512–530. doi:10.1080/00045608.2015.1018773

Luo, F., *et al.*, 2016. Explore spatiotemporal and demographic characteristics of human mobility via Twitter: a case study of Chicago. *Applied Geography*, 70, 11–25. doi:10.1016/j.apgeog.2016.03.001

Mahmud, J., Nichols, J., and Drews, C., 2012. Where is this Tweet from? Inferring home locations of Twitter users. *ICWSM*, 12, 511–514.

Mathew, W., Raposo, R., and Martins, B., 2012. Predicting future locations with hidden Markov models. *In: Proceedings of the 2012 ACM conference on ubiquitous computing*, 911–918. New York, NY: ACM.

Mennis, J. and Guo, D., 2009. Spatial data mining and geographic knowledge discovery—an introduction. *Computers, Environment and Urban Systems*, 33 (6), 403–408. doi:10.1016/j.compenvurbsys.2009.11.001

Moreira, G., Santos, M.Y., and Moura-Pires, J., 2013. SNN input parameters: how are they related? *In: Parallel and distributed systems (ICPADS), 2013 international conference on*, Seoul, South Korea, 492–497. Piscataway: IEEE.

Morstatter, F., *et al.*, 2013. Is the sample good enough? Comparing data from twitter's streaming api with twitter's firehose. *In: Proceedings of the 7th International Conference on Weblogs and Social Media*, ICWSM 2013, Jul 8 -11, Cambridge, MA, United States. Palo Alto: AAAI press.

Noulas, A., *et al.*, 2012a. A tale of many cities: universal patterns in human urban mobility. *PloS one*, 7 (5), e37027. doi:10.1371/journal.pone.0037027

Noulas, A., *et al.*, 2012b. Mining user mobility features for next place prediction in location-based services. *In: ed. Data mining (ICDM), 2012 IEEE 12th international conference on*, 10–13 December. Brussels, Belgium, 1038–1043. Piscataway: IEEE.

Parvez, A.W.M.M., 2012. Data set property based 'K'in VDBSCAN clustering algorithm. *World of Computer Science and Information Technology Journal (WCSIT)*, 2 (3), 115–119.

Rand, W.M., 1971. Objective criteria for the evaluation of clustering methods. *Journal of the American Statistical Association*, 66 (336), 846–850. doi:10.1080/01621459.1971.10482356

Richardson, D.B., *et al.*, 2013. Spatial turn in health research. *Science*, 339 (6126), 1390–1392. doi:10.1126/science.1232257

Santos, J.M. and Embrechts, M., 2009. On the use of the adjusted rand index as a metric for evaluating supervised classification. *In*: ed. *International conference on artificial neural networks*, 14–17 September. Limassol, Cyprus, 175–184. Berlin: Springer.

Shaw, S.-L., Tsou, M.-H., and Ye, X., 2016. Human dynamics in the mobile and big data era. *International Journal of Geographical Information Science*, 30 (9), 1687–1693. doi:10.1080/13658816.2016.1164317

Steiger, E., Albuquerque, J.P., and Zipf, A., 2015. An advanced systematic literature review on spatio-temporal analyses of Twitter data. *Transactions in GIS*, 19 (6), 809–834. doi:10.1111/tgis.12132

Walde, S.S.I., 2006. Experiments on the automatic induction of German semantic verb classes. *Computational Linguistics*, 32 (2), 159–194. doi:10.1162/coli.2006.32.2.159

Wang, S., Liu, Y., and Shen, B., 2016. MDBSCAN: multi-level density based spatial clustering of applications with noise. *In*: ed. *Proceedings of the 11th international knowledge management in organizations conference on the changing face of knowledge management impacting society*, 25–28 July. Hagen, Germany, 21. New York, NY: ACM.

Xu, Y., *et al.*, 2016. Another tale of two cities: understanding human activity space using actively tracked cellphone location data. *Annals of the American Association of Geographers*, 106 (2), 489–502.

Yeung, K.Y. and Ruzzo, W.L., 2001. Details of the adjusted rand index and clustering algorithms, supplement to the paper an empirical study on principal component analysis for clustering gene expression data. *Bioinformatics*, 17 (9), 763–774.

Yuan, J., Zheng, Y., and Xie, X., 2012. Discovering regions of different functions in a city using human mobility and POIs. *In*: ed. *Proceedings of the 18th ACM SIGKDD international conference on knowledge discovery and data mining*, 23–25 July. Edmonton, AB, Canada, 186–194. New York, NY: ACM.

Zandbergen, P.A., 2009. Accuracy of iPhone locations: A comparison of assisted GPS, WiFi and cellular positioning. *Transactions in GIS*, 13, 5–25. doi:10.1111/tgis.2009.13.issue-s1

Zhou, C., *et al.*, 2004. Discovering personal gazetteers: an interactive clustering approach. *In*: ed. *Proceedings of the 12th annual ACM international workshop on geographic information systems*, 12–13 November. Washington, DC, USA, 266–273. New York, NY: ACM.

Same space – different perspectives: comparative analysis of geographic context through sketch maps and spatial video geonarratives

Andrew Curtis, Jacqueline W. Curtis ⓘ, Jayakrishnan Ajayakumar ⓘ, Eric Jefferis and Susanne Mitchell ⓘ

ABSTRACT

The importance of including a contextual underpinning to the spatial analysis of social data is gaining traction in the spatial science community. The challenge, though, is how to capture these data in a rigorous manner that is translational. One method that has shown promise in achieving this aim is the spatial video geonarrative (SVG), and in this paper we pose questions that advance the science of geonarratives through a case study of criminal ex-offenders. Eleven ex-offenders provided sketch maps and SVGs identifying high-crime areas of their community. Wordmapper software was used to map and classify the SVG content; its spatial filter extension was used for hot spot mapping with statistical significance tested using Monte Carlo simulations. Then, each subject's sketch map and SVG were compared. Results reveal that SVGs consistently produce finer spatial-scale data and more locations of relevance than the sketch maps. SVGs also provide explanation of spatial-temporal processes and causal mechanisms linked to specific places, which are not evident in the sketch maps. SVG can be a rigorous translational method for collecting data on the geographic context of many phenomena. Therefore, this paper makes an important advance in understanding how environmentally immersive methods contribute to the understanding of geographic context.

Introduction

It is now widely accepted that official quantitative spatial data, such as commonly used in crime or health analyses, has limitations for the identification of both problems and the causal mechanisms to inform solutions. These data lack context, the insights of those who can best explain where and why these data points occur, what the implications are and how various social systems weave together through these locations. A major challenge lies in how to capture the geography of these types of insights. Many methods have been employed in this endeavor, and it is now appropriate to begin the process of assessing them in comparison to one another with the aim of identifying their contributions to understanding entrenched social problems. Therefore, the aim of this paper is to make progress in this area through comparison between a traditional and an emergent method in a case study with criminal ex-offenders.

A long-standing approach has been to use sketch maps to tap into people's knowledge, preference, perceptions and behaviors. More recently, different technologies have allowed for more sophisticated ways to capture even greater depth by having a participant interact with his/her environment and using its stimuli to trigger both spatial and social insights (eg Kwan and Ding 2008, Evans and Jones 2011, Mennis *et al.* 2013). While initial results have been exciting, there remains much to do in terms of evaluating how these stimuli, spatial recall, experiences and the methods themselves influence our understanding of geographic context. We also need to see how the more logistically challenging and expensive methods compare against those that are cheaper, quicker and more widely accessible. Therefore, in this study, we collect spatial video geonarratives (SVGs) from 11 ex-offenders for one neighborhood and compare results with the sketch maps created by the same participants during the same time. These results are compared against police calls for service, which has previously been used to create hot spots through keyword selection. Then, in the discussion, we focus more on the content of the narratives in search for an explanation of where we find variations between these different data layers and to what degree is our search for more meaningful geographic context furthered using this method.

Geospatial methods for geographic context

This study responds to the call for further research into methods for mitigating the uncertain geographic context problem (Kwan 2012a, 2012b). In particular, it builds on existing work on geonarratives (Kwan and Ding 2008) and their potential to address the recognition that data relevant to solving real-world problems must reflect the real-world spatial-temporal dynamism in people's daily lives (Kwan 2013). To achieve this aim, this study begins with a traditional approach, sketch mapping. This long-standing and widely accessible technique for accessing the geography of participants' knowledge, preferences, perceptions and behaviors can be implemented in a number of ways, though often by giving people a base map with some basic spatial reference information (eg roads) and then asking them to mark locations in response to specific questions (Curtis 2016). Although sketch mapping is grounded in work from the 1960s and the 1970s (Lynch 1960, Gould and White 1968, Gould and White 1974, Downs and Stea 1974), it is experiencing a resurgence, in part through integration with GIS. Recent examples include applications toward understanding environmental risks (O'Neill *et al.* 2015), public health (Beyer *et al.* 2010, Manton *et al.* 2016) and most prominently in crime/fear of crime (Doran and Lees 2005, Lopez and Lukinbeal 2010, Doran and Burgess 2011, Curtis *et al.* 2014). Although sketch maps make a number of contributions such as opening opportunities for countermapping perspectives, or the generation of detailed spatial information of individuals, and facilitation of data interpretation (Boschmann and Cubbon 2014, p. 236), little is known about how the results from this method compare to other approaches aimed at understanding geographic context.

Moving from more widely accessible methods, such as sketch mapping, to those that leverage emergent geospatial technology, an advance that is showing promise toward understanding geographic context is the idea of geographically placed narratives (commonly, but not always, termed 'geonarrative' or 'geo-narrative').[1] This approach seeks to integrate narrative analysis with a geographic frame, specifically through modern geospatial technology, for the locations of features, events, behaviors, etc. (Kwan and Ding 2008). Foundational work in this area has taken different forms, such as combining participant sketch maps with travel diaries and audio-recorded narratives of Muslim women (Kwan and Ding 2008) to risk in activity spaces of youth identified through Ecological Interviews (Mennis et al. 2013). Building on the insight that these existing methods have contributed to spatial and even spatial-temporal processes, researchers are now moving to embrace the idea of geographic embeddedness or emplacement, collecting data with participants as they move or are moved (eg walking, driving) through a particular place or space (Elwood and Martin 2000, Anderson and Jones 2009). For example, Jones and colleagues (2008), Evans and Jones (2011) and Bergeron et al. (2014) have examined issues surrounding planning, redevelopment and place attachment. In addition, Bell and colleagues (2015, 2017) have utilized 'go-along' interviews to identify activity spaces and understand green space and blue space experiences. Others have utilized such methods to examine environmental activism (Anderson 2004), health–place relationships (Carpiano 2009), urban renewal and aging (Lager et al. 2013), and even walking itself, especially among those with disabilities (Butler and Derrett 2014). In particular, the work of Evans and Jones (2011) is germane to understanding the impact of geographic embeddedness on interview methods. They generated 'spatial transcripts' linking GPS to audio recording in walked interviews and compared the content to sedentary interviews and those who participated in both types of interviews. The spatial transcripts were overlaid in a GIS to visualize multiple participant input over places. Their results show that in comparison to sedentary interviews, interviews conducted while walking through the environment are quite different, with the embedded interviews yielding more geographically specific features as well as more explanation.

In response to the growth of this vein of methodological inquiry, it is important to be cognizant of the thoughtful critiques that are emerging. These include probing data accuracy (Merriman 2014), participant safety (Carpiano 2009) and participant exclusion. More recently, Warren (2017) provided evidence of the need to approach this method more critically as it relates to a range of vulnerable populations and the need to pluralize the walking interview using a case study of Muslim women. Her study points to the need for reflection on who is likely to participate in such an interview given domestic and other caregiving responsibilities of women; likewise, such reasons may also influence which researchers can use this approach. In sum, not everyone wants to be visible in her/his community for a variety of reasons, nor will all feel comfortable leading such an interview – determining the route and openly sharing thoughts inspired by the environment, which can be based on a range of social and cultural reasons.

While we recognize and appreciate the potential limitations of embedded interviews, for this study we believe our participants provide a suitable set of subjects to consider the complex interactions of crime in their home neighborhood of a medium-sized midwestern city through a further methodological advance on embedded techniques, the SVG.

SVG shares the intention of many extant sketch map applications, but is also closely aligned with the emergent work on geographically embedded mobility. Specifically, it is an outgrowth from the geonarrative/geo-narrative through the use of GPS-encoded audio and video data. SVG has been used by the authors in a variety of different environments to capture local context (Curtis et al. 2015, 2016, Krystosik et al. 2017, Schuch et al. 2017). Participants are professionals, residents or others with an intimate understanding of the environment being traversed. Typically, a car with between two and five spatially enabled cameras captures the passing environment while the partici-pant narrates his/her comments responding to various stimulation of the senses. While visual cues are the most frequent conversation generator, sounds and smells and even 'feelings' have also led to a topic being discussed. The content of this narrative covers distant past, recent and current experiences, while in some cases the expertise of the individual is used to interpret what is witnessed. Following from protocols established in prior research, content from these narratives is assigned to one of three categories: spatially specific, where precise locations are pointed out and described; spatially fuzzy, meaning the comment is directed toward the general area; and spatially inspired, where the comment is not spatial (though it may mention other spatial locations) (Curtis et al. 2015, 2016).

With implementation of these methods and their variants toward the objective of improving understanding of geographic context, it is appropriate to begin the process of assessing them in comparison to one another with the intention of identifying their strengths and limitations. Therefore, the aim of this paper is to make progress in this area through comparison between the traditional sketch map and the emergent SVG in a case study with criminal ex-offenders. Specifically, we ask the following questions: 1) What is the geography of high-crime areas identified by the sketch maps in comparison to the SVGs? 2) How do the descriptive attributes of these places derived from sketch maps and SVGs compare? 3) What do each of these datasets contribute to understanding the geographic context of crime in this community? Answers to these questions are then more fully considered using the richness of the narratives to tease apart how sensory stimulus and personal experience result in a contextualized map.

Materials & methods

This study reports on one aspect of a larger investigation on Offender Decision-Making using Geo-Narratives, which aims to create new understanding of the geographic context of crime covering many areas of the city. All participants were recruited in connection with a faith-based nonprofit organization in this community and were selected based on their local knowledge of crime in the area. The estab-lished relationship between researchers and this organization and the snowballing approach employed led to a relatively easy recruitment process. They were told the intention of the study was to assess how the neighborhood environment contributed to crime. They were compensated with twenty-dollar gift cards, and this compensa-tory payment was noted at the outset. Informed consent was obtained by all in accordance with Kent State University Institutional Review Board approval for this project (#13-522).[2]

Participants

Sketch maps followed by an SVG were collected from 11 ex-offenders over a short period of time (July–September 2015) for the same neighborhood. Each of the participants had been charged for drug dealing, 10 out of the 11 had been involved in prostitution and nine out of the 11 had committed burglary. Other common activities in which they had personal experience include theft, robbery, shootings and gang involvement (Table 1). The SVGs would reveal that these crimes/activities could actually be considered symptoms of drug addiction as each was a recovering addict.

Data collection

Data collection occurred in three stages: demographic survey, sketch map and SVG. The survey was composed of four pages (Supplementary Material – Appendix A). Participants were asked to fill out seven background questions that documented their age, sex, race, ethnicity, nativity, home ownership status in the community and educational attainment. Then they completed a sketch map. Each participant was given a legal sized map of the study area at a scale of 1:10,000 so that road names were clearly visible. Participants were instructed to use three colors for their markings: red for high-crime locations, green for crime-free areas and blue for locations where they were likely to spend time (this originally was a yellow marker, but blue was more visible, so this color was utilized instead). The participants were also asked to number each marked area and on the back of the map, write down the prevalent crimes for the labeled locations. A digital copy was made of each paper sketch map before being georegistered using TIGER Line road data for the study area in ArcGIS 10.4. The markings of each participant were digitized as polygons from the georegistered maps and characteristics written about each location added in the attribute table (Curtis *et al.* 2014).

Table 1. Participant demographics.

ID	Connection	Years	Age	Gender	Race	Criminal offenses
B070315	RESIDENT	<1	47	F	WH	Drug dealing, Burglary, Gang Activity, Car Theft, Prostitution, Robbery, Drug use
B072115	RESIDENT	<1	29	F	PI	Drug dealing, Burglary, Gang Activity, Prostitution, Fighting, Robbery, Drug use
072115	RESIDENT	10+	39	F	WH	Drug dealing, Burglary, Gang Activity, Car Theft, Prostitution, Fighting, Robbery, Drug use, Shootings
B072315	RESIDENT	10+	55	F	BL	Drug Dealing, Prostitution, Fighting, Robbery, Drug Use, Shootings
073015	VOLUNTEER	10+	35	F	WH	Drug Dealing, Burglary, Prostitution, Fighting, Drug Use, Shootings
B073015	RESIDENT	<1	47	F	WH	Drug Dealing, Burglary, Gang Activity, Car Theft, Prostitution, Robbery, Drug Use
080715	RESIDENT	1–3	41	F	WH	Drug Dealing, Burglary, Car Theft, Prostitution, Fighting, Drug Use
B090815	RESIDENT	10+	32	F	WH	Drug Dealing, Gang Activity, Prostitution, Fighting, Robbery, Drug Use (former dealer, heroin addict)
070915	RESIDENT	7–9	40	M	WH	Drug Dealing, Burglary, Prostitution, Fighting, Robbery, Drug Use, Shootings, Rape, Gambling, Sex Trafficking
072315	RESIDENT	<1	29	F	PI	Drug Dealing, Burglary, Gang Activity, Prostitution, Fighting, Robbery, Drug Use Shootings
101615	VOLUNTEER	10+	45	F	W	Drug Dealing, Burglary, Prostitution, Drug Use

After completion of the survey and sketch map, the interviewer asked if the participant would feel comfortable taking them to the high-crime areas/places that s/he had identified on the map. At this point, the SVG ride began with one researcher driving while the participant sat in the front seat and at least one other researcher sat in the back seat overseeing equipment and helping with questions. At least one female researcher was always in the car. The interviewer would ask some general opening questions, such as, *we would like you to describe the areas identified on the maps, and whatever else you see or comes to mind during the drive*. The participant was then left to dictate the route. As much as possible, the interviewer restrained himself/herself to clarifying questions, prompting questions in moments of prolonged silence, directional clarification and a few standard questions such as *What do you think could be done to improve this neighborhood?* The car consisted of four spatial video cameras, two on either side, and at least two audio recorders. Once the ride was completed, all videos were downloaded and checked for image quality, audio clarity and GPS accuracy. These, in addition to the mapped route and overall camera performance, were turned into a standardized ride metadata sheet. The first audible word found on the audio and video media devices (which also had to have a valid GPS coordinate) was extracted to provide the linking needed for mapping.

Analysis

Owing to the variety and nature of data collected and the objectives of this study, both quantitative and qualitative analyses were utilized. First, for the sketch maps, a grid was constructed for the study area (50 m x 50 m), and then a spatial join created a count of participant markings intersecting each grid cell. Heat maps were created based on these counts to aggregate individual participants' knowledge about areas of high crime into a collective picture of the study area (Curtis *et al.* 2014).

The SVG audio was transcribed with time stamps inserted before each comment, and this was merged with the GPS path using Wordmapper 2^3 software. Each narrative was read for themes, both topically (eg related to prostitution) and spatially. Wordmapper 2 was used to assign each comment as spatially specific (positive and negative), spatially fuzzy (positive and negative) and spatially inspired (positive and negative). For a comment to be spatially specific, the participant would have to identify or draw attention to a passing place, building or person. For spatially fuzzy, the comment would still be tied to the general proximity, for example a comment about the blight around these streets, with the remaining comments falling into the inspired category. The spatial output from Wordmapper 2 (Google Earth KML and comments shapefile) had this spatiality assignment identified as a colored pin (KML) or as a code in the attribute table (shapefile). For the purposes of this paper, all spatially specific comments about criminal activity were overlaid on the sketch map outputs.

Police calls for service had also been acquired as an alternative to more traditional crime-related data. A key word query for either 'drug related' or 'violence related' calls in effect generated a numerator file for a spatial filter analysis,[4] while the denominator was all calls. This analysis was for comparative purposes in terms of how *spatially* aligned were the ex-offender insights with official police data.

To answer the research questions, several additional manipulations occurred of the previously described data and analytical outputs. First, for each participant, high-crime areas identified through their sketch maps were overlaid with their SVG ride path (Supplementary Material – Appendix C), and the percentage of the route that intersected with the sketch map area was calculated (with the extent of the base map used in the sketch map digitized as a polygon). A spatial query then selected all SVG coordinates that intersected this base map polygon. This procedure was used to compare the geographic extent of high-crime areas as identified by the participants in the sketch map and the SVG. Then, in order to compare the spatial pattern of locations identified in the sketch maps and in the SVGs, all spatially specific crime locations were queried from the Wordmapper 2 shapefile output and buffered at 25 m, 50 m and 100 m. Three buffer sizes were used to provide different geographic scales around each identified coordinate location as the path GPS did not always sit on top of, or immediately proximate to, the location being described. These buffers provide a counter for such artificial precision. Again, a spatial query was performed to identify sketch map polygons that intersected the buffers. As the SVG was driven and limited to roads, a second spatial query was performed to identify polygons that were within 50 m of a buffered area, specifically to capture places that are not immediately proximate to the road (eg parks, apartment complexes, etc.). Finally, comparing and contrasting the attributes assigned to these spaces were accomplished through a count of the number and documentation of type of characteristics for each identified location.

Results

Based on their local knowledge, the 11 participants marked 139 high-crime locations on sketch maps. To visualize these data, heat maps were constructed by overlaying the digitized sketch map markings (Curtis *et al*. 2014). A 50 m × 50 m grid was applied to enable aggregation of individual markings and create maps showing the continuum of spaces of concern, from where few participants marked an area to where many identified the same place (Figure 1). Three areas of some level of consensus are identified as Hot Spots A, B and C.

Although this approach can identify some common areas of high crime, examination of the size and shape of the participants' markings indicates varying geographies from specific locations along a street segment or within a city block to larger areas comprising several blocks. The descriptions of crime identified by the participants also vary within each of the three identified hot spots (Supplementary Material – Appendix D).

Note that in Hot Spot A, one participant writes 'drugs', while for another it is 'drug houses, drug sales' and a third states 'drugs (meth labs, heroin, drug dealing)', so in addition to what attributes are assigned to any particular area, there is also variation in the descriptions provided even about the same activity.

In comparison, the SVGs generate far more detail about places, events and times. To provide a direct comparison with the sketch maps, only the spatially specific negative comments were used. The spatially specific locations for each participant were extracted as a separate shapefile and buffered to three distances: 25, 50 and 100 m. There is a range of between 3 and 16 spatially negative buffers at the smallest buffer size[5] (containing between 5 and 91 comments) across the participants. Subject 080715 had

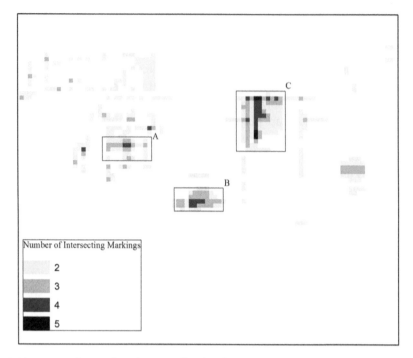

Figure 1. Hot spots of crime based on ex-offender sketch maps.

both the shortest ride and the fewest spatially precise mentions, with most comments being more general about the neighborhood. Interestingly, the number of buffers and the number of spatially specific mentions do not vary consistently. For example, there can be more detailed description of a smaller area or single descriptions of more scattered locations.

As a way to display the frequency of geographies occurring across all SVGs, a grid was overlaid on the map that would also be used for the spatial filter analysis. If a grid node fell inside a buffer, it received a count, the sum of which gave a distribution of how frequently that area was mentioned in the SVGs. These frequencies are displayed in Figure 2, with the highest number of overlays being 5, meaning five different test subjects mentioned at least one spatially significant negative location within 100 m of that grid point. This map is interesting in that it shows four locations (A, B, C and D) where there appears to be consensus among multiple participants as to clearly identifiable problem locations, with A revealing the highest concentration of consensus on problematic spaces and places.

To explore how these spatially specific SVGs compare with official data and a more aggregated way of working with SVG data, the spatially specific buffers were overlaid onto spatial filter output surface for calls for service. For the purposes of this paper, $p = 0.05$ for violence and drug calls for service spatial filter outputs are displayed. Figure 3 displays an area worthy of further discussion identified in Figure 2 as 'A'. Only the inner polygon of the buffer has been shaded so that multiple overlays can be seen together. These maps also show the results of the 2015 $p = 0.05$ spatial filter output. In Figure 3 (location A), the top left inverted 'L' displays three streets that

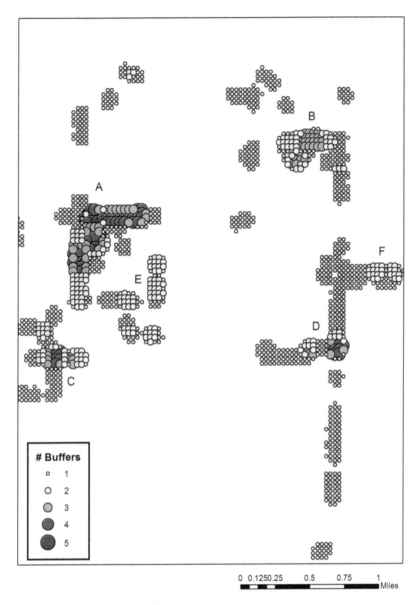

Figure 2. Frequency of spatially specific mentions.

received a high number of spatially precise negative mentions from six different SVGs, with three others being in close proximity. This same area also had statistically significant filter outputs for both drug and (especially) violence calls for service. This map also reveals other interesting patterns, for example the central area that has statistically significant filter output for violence-related calls for service, but no spatially specific mentions made in the vicinity by the ex-offenders. Location B in Figure 2 is also notable in that it has four participants' SVGs identifying spatially precise locations, and yet no statistically significant calls for service occur in this vicinity.

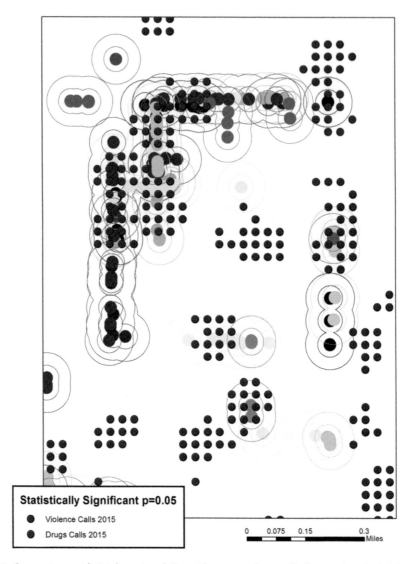

Figure 3. Comparison with 'Violence' and 'Drug' hot spots from calls for service. Dark blue points represent calls for service regarding violence, dark red points represent calls for service related to drugs, and all other points and buffers represent locations of SVG mentions.

Moving from separate to comparative investigation of sketch maps and SVGs, our first question is about the similarities and differences of their geography, both the geographic extent and then the places and patterns within these areas. Starting with the geographic extent, Table 2 provides a description of the continuity and discontinuity of space resulting from each method.[6]

The percentage of SVG routes directed by participants that are located off the researcher-defined sketch map area ranges from 0 to 59%, with a median of 32%. This means that there is a notable area of the neighborhood that is meaningful to the participants that was not even made available for them to include in the sketch map, despite researchers' due diligence to use an appropriate geographic extent in the base

Table 2. Comparison of sketch map to SVG data: continuity/discontinuity of geographic extent identified by participants.

ID	SVG PATH (coordinate points)	OFF THE MAP	ON THE MAP
B070315	5756	1821 (31.64%)	3935 (68.36%)
B072115	4484	1615 (36.02%)	2869 (63.98%)
072115	7733	4622 (59.77%)	3111 (40.23%)
B072315	4415	1175 (26.61%)	3240 (73.39%)
073015	6961	2292 (32.93%)	4669 (67.07%)
B073015	2860	0 (0.00%)	2860 (100.00%)
080715	4054	1308 (32.26%)	2746 (67.74%)
B090815	3578	1730 (48.35%)	1848 (51.65%)
070915	7116	2232 (31.37%)	4884 (68.63%)
072315	6347	3603 (56.77%)	2744 (43.23%)
101615	5917	2534 (42.83%)	3383 (57.17%)

map. This result points to the limitation of using paper sketch maps as their extent is static, but it also raises the importance of testing more spatially dynamic data collection tools such as softGIS (eg Kyttä et al. 2013) and Maptionnaire (https://maptionnaire.com/) as well as other options that use digital base maps on tablet technology (eg Google Earth/Maps as a base map for sketch map data collection).

Moving from comparison of the overall geographic extent defined by the participants, the following section focuses on the geography of high-crime locations identified within the study area. All participant SVG paths intersected at least one of the high-crime locations each identified through their sketch map, though approximately half of the sketch map locations do not intersect or are not even proximate (within 50 m) to the spatially specific places identified in the SVG. This suggests that even when participants directed the SVG route through the same geographic areas identified on the sketch map, they were not consistent in identifying the same locations.

A further notable difference between results from the two methods is in what type of events was described and in what level of detail. The participants produced a more nuanced and deeper understanding of the neighborhood through the SVG. Comparison of the fourth and fifth columns in Table 3, which are the descriptions of high-crime locations in the sketch map and in the SVG, reveals stark differences.

First, the attributes assigned to places in the sketch map are often noted as general categories of criminal offenses, as opposed to descriptions provided in the SVGs that are more detailed. Then, the number of attributes identified in sketch maps and SVGs differs, with sketch maps producing from as little as three to a high of 14, while SVGs consistently generated more. Finally, the types of attributes assigned to places also differ. In the sketch map, the descriptions are primarily types of crimes, while in the SVG, they are about types of crimes, as well as specific people, places and spatial-temporal relationships related to the crimes. These observations are qualified by the fact that the sketch maps took no more than 10 min to complete and the SVGs often lasted up to 1 hour, so with more time, there is more opportunity for locating and describing places and processes. However, we suggest that even with equal amounts of time, SVGs would still lead to better data on geographic context. This is a question that should be investigated in future research.

Table 3. Comparison of sketch map to SVG data: number of places identified and excerpts of attributes for these locations. In SVG NUMBER, the first value is the number of buffered areas using the largest buffer and the value in parentheses is the number of buffered areas using the smallest buffer.

ID	SKETCH NUMBER	SVG NUMBER	SKETCH DESCRIPTIONS	SVG DESCRIPTIONS
B070315	3	15 (30)	1: DRUGS, GANGS 2: THEFT, LOITERING 3: DRUGS, VIOLENCE, BURGLARY-ROBBERY, AUTO THEFT	1: DRUGS – CRACK 2: DRUGS – CRACK 2&3: DRUGS – CRACK 3: BUILDING ON THIS STREET SEGMENT WHERE DRUGS ARE SOLD A: A STREET THAT HAD HIGH CRIME, BUT HAS IMPROVED B: A LOCATION (NAME REMOVED) IN THE STREET SEGMENT THAT HAD HIGH CRIME, BUT HAS IMPROVED C: STREET THAT WAS DANGEROUS AT NIGHT D: BOUNDARY/TRANSITION ZONE BETWEEN LOW- AND HIGH-CRIME AREAS E: ROUTE FROM HOME TO ACQUIRE DRUGS F: STREET WITH PREVALENCE OF THEFT, AUTO THEFT, ROBBERY, ASSAULT G: HOUSE WHERE DRUGS COULD BE PURCHASED H: BOARDED HOUSE THAT IS A CRACK HOUSE I: STORE WHERE DRUG DEALERS WERE PRESENT J: APARTMENTS WHERE CRACK IS SOLD
B072115	6	8 (31)	PROSTITUTION GANGS SHOOTINGS BURGLARY/ROBBERY	TALKING ABOUT SKETCH MAP: METH LABS, BURGLARY – A LOCATION IDENTIFIED ON THE MAP A LOCATION ON THE MAP – PROSTITUTION A: SHOOTINGS, CRACK AND DOPE USE B: PROSTITUTION C: BRIDGE – FOR DRUG USE BECAUSE THESE IS A 'HIDING SPOT IN THE BUSHES' D: DRUG HOUSES E: 'WHERE PEOPLE MEET TO GET DRUGS' F: APARTMENTS WITH A METH LAB

(Continued)

Table 3. (Continued).

ID	SKETCH NUMBER	SVG NUMBER	SKETCH DESCRIPTIONS	SVG DESCRIPTIONS
072115	6	8 (28)	1: MURDER 2: THEFT 3: PROSTITUTION 4: GANGS 5: DRUGS 6: ROBBERY 7: ALL CRIME (NUMBERS ARE NOT LABELED ON THE MAP)	A HOUSE WHERE DOPE WAS COOKED BOARDED DOORS AS A SIGN THE METH IS COOKED THERE OR IT WAS A RAIDED CRACK HOUSE BRIDGE FOR DRUG USE BECAUSE 'WE COULD JUST THROW OUR SHIT OVER THE SIDES OF THE HIGHWAY' LOCATION OF SHOOTINGS STREET WHERE IN THE PAST THE SMELL FROM DOPE COOKING WAS PREVALENT HOUSE WHERE SPEED WAS BEING MADE – BASED ON SMELL STREET WITH CRACK HOUSES AND PROSTITUTION DENS STREET SEGMENT KNOWN AS 'THE HOE STROLL' ANOTHER STREET KNOWN AS 'THE HOE STROLL' LOCATION OF A DRUG SALE HOUSE WHERE HEROIN IS SOLD CORNER WHERE A CHILD WAS MURDERED SCRAP YARD WHERE STOLEN METAL IS SOLD FOR MONEY TO BUY DRUGS ADULT CINEMA WHERE THERE ARE PROSTITUTES AREA WITH HOUSES THAT ARE BROTHELS A BUILDING THAT IS 'ONE OF THE WORST PLACES FOR DRUGS' AREA WITH PREVALENCE OF POT, ACID, COCAINE STRIP CLUB WHERE MAJORITY OF DANCERS ARE ADDICTED TO DRUGS OR ALCOHOL PLASMA – WHERE PEOPLE GIVE PLASMA FOR MONEY TO SPEND ON DRUGS APARTMENT THAT IS 'DOPE INFESTED'
B072315	1	9 (60)	MURDER, GANGS, DRUGS, PROSTITUTION, HOMELESS, BURGLARY, FIGHTING, RAPE	GROUP OF BUILDINGS WHERE PEOPLE RENT ROOMS THAT ARE 'DOPE INFESTED' STRIP CLUB AND SURROUNDING AREA THAT IS DANGEROUS APARTMENT BUILDING WITH DRUGS A DOPE HOUSE STREET THAT IS A 'HOE STROLL' WITH DRUG SALES AND USE GROUP OF APARTMENTS ON TOP OF BUSINESSES WHERE DRUGS ARE PRESENT A DOPE HOUSE WHERE A MURDER OCCURRED PARK WITH DRUG DEALERS BUILDING THAT IS 'DOPE INFESTED' AREA WITH METH HOUSES APARTMENT WHERE DOPE IS SOLD ABANDONED HOUSE BEING USED TO COOK METH AREA OF DOPE HOUSES A DOPE HOUSE

(Continued)

Table 3. (Continued).

ID	SKETCH NUMBER	SVG NUMBER	SKETCH DESCRIPTIONS	SVG DESCRIPTIONS
073015	4	8 (56)	1: DRUGS, PROSTITUTION, SHOOTINGS, ASSAULTS 2: DRUGS 3: DRUGS, PROSTITUTION, SHOOTINGS, ASSAULTS, BURGLARIES 4: DRUGS, SHOOTINGS, PROSTITUTION, ASSAULTS	LOCATION WHERE S/HE PARKED A CAMPER VAN TO COOK METH ABANDONED HOUSE WHERE METH IS COOKED HOTEL WHERE DRUGS ARE SOLD AND THERE IS A STRIP CLUB, DRUG USE, PROSTITUTION AND 'A LOT OF SEX OFFENDERS LIVE THERE' STREET THAT IS NOT SAFE TO WALK IN DAY OR NIGHT DUE TO ROBBERY VACANT HOUSE WHERE S/HE HAS SMELLED DRUGS COOKING AND THE GARAGE NEXT TO IT AND A NEARBY ALLEY GROUP OF HOUSES WITH METH AND WHERE A SHOOTING OCCURRED STREET WITH PROSTITUTION STREET WITH DRUGS GROUP OF STREET WHERE CRACK IS AVAILABLE APARTMENT WITH PROSTITUTION APARTMENT WITH DRUG DEALERS A STREET THAT IS A 'HOE STROLL'
B073015	7	8 (36)	DRUG DEALING, SHOOTINGS, BEATING, STABBINGS, PROSTITUTION	HOUSES WHERE DRUG DEALERS LIVE SHOOTING AROUND THE HOUSES WHERE DRUG DEALERS LIVE A FIELD WHERE A GIRL WAS STABBED A STREET WHERE PROSTITUTES WALK AND DRUGS ARE USED BUILDING ON A CORNER WHERE HEROIN IS SOLD HOUSE WITH BULLET HOLES LOCATION WHERE ASSAULT WITH HAMMER OCCURRED AND IN THE SAME AREA A BLACK WOMAN MURDERED A WHITE MAN AND A SHOOTING. ALL DRUG-RELATED PARK – SPECIFICALLY AT THE BASKETBALL HOOP DRUGS ARE SOLD ALLEYS WHERE DRUG DEALS OCCUR DRUG HOUSES STREET WHERE DRUGS WERE BEING USED IN ABANDONED HOUSES, BUT HOUSES DEMOLISHED, SO THEY NOW USE A NEARBY FIELD AND GARAGES BAR SHUT DOWN DUE TO SHOOTINGS CORNER WHERE DRUGS ARE SOLD
080715	4	3 (5)	THEFT, DRUGS, FIGHTING, WEAPONS	AREA WHERE DRUGS ARE SOLD OPENLY STREET WITH PREVALENCE OF CRACK AREA AROUND A BAR WHERE GUNSHOTS WERE OFTEN HEARD STREET WITH DRUG DEALERS RAILROAD TRACKS WHERE A WOMAN WAS BEATEN

(Continued)

Table 3. (Continued).

ID	SKETCH NUMBER	SVG NUMBER	SKETCH DESCRIPTIONS	SVG DESCRIPTIONS
B090815	6	5 (17)	1: ...ARE BAD DRUG AREAS 2: ...BIKER CLUB AREAS 3: ...PROSTITUTION AND DRUG AREAS	DRUG HOUSE WHERE HEROIN IS SOLD HOUSE THAT LOOKS NICE, BUT 'LADY SELLS PILLS' TWO HOUSES WITH DRUG ACTIVITY STREET WITH PROSTITUTION APARTMENT COMPLEX WITH PROSTITUTION, DRUGS AND CRIMES AREA WITH DRUG ACTIVITY BIKE CLUB
070915	95	16 (66)	1: CRACK, WEED, PROSTITUTION; PAST X ROAD – HEROIN 2: THEFT, BREAKING & ENTERING, MUGGING, WEED, DOMESTIC DISTURBANCES 3: DRUG HOUSES, DRUG SALES ON CORNERS, GAMBLING, RAPE – DRUG RELATED AND LOCATION BASED	PARK THAT IS UNSAFE AT NIGHT GAS STATION WHERE DRUGS ARE SOLD AT NIGHT STRIP BARS (AND ONE IN PARTICULAR) WHERE HEROIN IS SOLD STREET SEGMENT WITH MURDERS AND MUGGINGS SET OF STREETS WITH DRUG ACTIVITY AT NIGHT APARTMENT COMPLEX WITH DRUG ACTIVITY APARTMENTS THAT ARE SECLUDED AND THEREFORE ARE A LOCATION FOR MURDERS SET OF STREET WITH HIGH CRIME A PARTICULAR STREET WITH HIGH CRIME APARTMENT COMPLEX WITH WEED AND HEROIN ACTIVITY A STREET SEGMENT WHERE DEALERS SELL HEROIN, ECSTASY, XANEX AND WEED AREA AROUND A BRIDGE WITH PROSTITUTION AND CRACK HOUSES AREA WITH CRACK, HEROIN, WEED AND PROSTITUTION MOTEL WHERE PROSTITUTION OCCURS FOR CRACK A STREET THAT IS KNOWN AS A 'CRACK WALK OF FAME' AND A BUSH AND HOUSE WHERE THE DRUGS ARE TAKEN HOUSES WHERE THERE IS DRUG ACTIVITY AND SHOOTINGS APARTMENTS KNOWN FOR HIGH CRIME – ALL CRIMES A SPECIFIC BUILDING IN THIS APARTMENT COMPLEX KNOWN FOR HEROIN, MURDERS, SHOOTINGS AREA OF CONNECTION BETWEEN APARTMENT COMPLEXES WHERE SUSPECTS CAN ESCAPE THE POLICE HOUSES WITH METH ADDICTS CHECK CASHING BUSINESS WHERE DRUG ADDICTS SELL STOLEN GOODS FOR DRUG MONEY, ESPECIALLY HEROIN ADDICTS

(Continued)

Table 3. (Continued).

ID	SKETCH NUMBER	SVG NUMBER	SKETCH DESCRIPTIONS	SVG DESCRIPTIONS
072315	3	9 (33)	DRUG ACTIVITY (METH LABS, HEROIN, DRUG DEALING) BURGLARY ROBBERY	A STORE WHERE DRUG DEALS OCCUR AREA WITH PROSTITUTION AREA WITH PROSTITUTION, DRUG DEALING, METH LABS, CRACK, HEROIN, 'YOU NAME IT, YOU WANT TO FIND IT, IT'S HERE' AN AREA THAT IS DANGEROUS AT NIGHT FOR BEING RAPED; ALSO DRIVE-BY SHOOTING IN THIS AREA A PROSTITUTE ABANDONED FACTORIES USED BY THE HOMELESS 'THERE'S A LOT OF ASS WHOOPINGS ON THAT CORNER' HIGH SCHOOL WHERE 'KIDS WILL WAIT FOR DRUG DEALER OUT TO THE LEFT' A DRUG RUNNER
101615	7	11 (31)	DRUG TRAFFICKING, PROSTITUTION, BURGLARY, DRUG USE, FIGHTS	COLLEGE HOUSING WITH DRINKING AND DRUGS LOCATION WITH DRUG ACTIVITY HOUSES WITH SIGNS OF METH USE AREA WHERE CRACK IS AVAILABLE PARK WITH DRUG USE AT NIGHT STREET WITH PROSTITUTION BARS AND ONE IN PARTICULAR WITH DRUGS

Discussion

Results from the SVGs reveal that not only do the participants' geographic contexts vary across space and over time, even in a small study area, but the contextual influences exhibit dynamism as well. There are many moving parts rather than a bounded neighborhood area serving as a container for exposures (eg drug sales, pimps) and the mitigating forces that aim to push back (eg churches, rehabilitation centers). This population of criminal ex-offenders, all recovering drug addicts, have been highly mobile in their past (renting, often not at the home address when seeking/using drugs, going in/out of rehabilitation) and still exhibit residential mobility, though less so in recovery. The extreme dynamism of their contextual influences related to the drug–crime nexus was also revealed through the SVGs. Kwan (2012b, 2013) has highlighted several examples (eg air pollution, neighborhood change) where contextual influences exhibit variability even over hours, the course of a day, etc. This study contributes an additional example with a topic that is intentionally covert and where the geographic context can be momentary and not easily foreseen. For example, the drug distribution sources change (different cities – Detroit, Cleveland, etc.; different organizations managing distribution), as do the drugs of choice (from prescribed opioids to heroin) to the exact composition of the drugs (eg introduction of fentanyl, carfentanil, etc.), to police interventions, all of which change the geographic context of drug-seeking behavior at multiple scales. Unlike previous areas of study that reveal the dynamism of geographic context, the drug–crime nexus changes fast, often violently, and is covert by design. It also is a major contributor to some of the most pressing and entrenched social, economic and public health problems in the United States. It is not well-suited to a clear, observable geographic and temporal extent, and therefore existing geospatial methods are incomplete in producing actionable knowledge to inform intervention.

This study suggests that SVGs may offer a contribution to advancing understanding of the geographic context of this problem. Specifically, this paper examined results from two methods for identifying geographic context: sketch maps and SVGs. It was reassuring to find that one area (C and A in Figures 1 and 3, respectively) was consistently identified in both the aggregate sketch maps and SVG maps (Figure 4); it was also revealed through the spatial filter analysis of calls for service. This suggests that, perhaps for the highest crime areas, there is general spatial consensus across traditional and nontraditional forms of data collection. However, while there is some spatial agreement, there is considerable variation in *what* is conveyed. In addition to different characteristics reported within the SVGs, in the sketch maps it became evident that some detail is lost in the translation from their mind to the map. For example:

> "can I say like beatings? Like when the guy beat the guy with the hammer in the face..." and then "...ah prostitution, I don't spell good... and that's it, now there's a girl that has AIDS and she spreads that around over here... so I'm going to say prostitution."

None of this detail was provided on the sketch map, only in her/his audio of that process. This does not mean we should abandon the use of sketch maps, far from it. If we consider the methods for representing geographic context to be a continuum, sketch maps provide a widely accessible tool that will still enrich or complement more

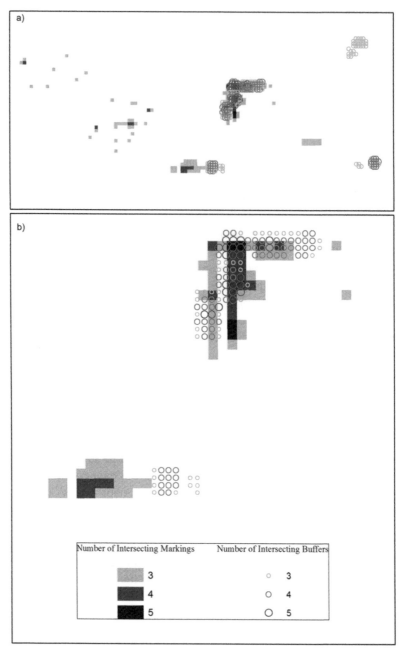

Figure 4. Overlay of crime hot spots from sketch maps and spatial video geonarratives: a) study area and b) zoom-in to hot spots.

traditional data. Sketch maps have low technology requirements (eg paper, markers, GIS), they can be administered relatively quickly (eg 10 min) and therefore are appropriate when obtaining the perspective of a large number of participants. Given these results, sketch mapping would benefit from thoughtful reconceptualization with a) a dynamic base map (tablet) and b) audio recording of the narrative while sketch mapping.

By comparison, the SVG, while being more logistically intensive, identifies more spatially specific locations and with much greater depth. This result is due to some combination of reasons that should be isolated and studied individually in future research. First, it could be that the sketch maps require a level of comfort in their use, drawing upon participants' map literacy and graphicacy and therefore potentially limiting the amount and type of data acquired through this method. In contrast, SVGs do not pose such a barrier. Second, it is possible that the sketch maps primed the participants and therefore they were thinking more about the relevant issues in their neighborhood, which contributed to a more detailed narrative. Furthermore, sketch maps took considerably less time to complete, while the SVGs look longer. This difference was not an intention of the study design, but rather this is how the methods organically played out in reality. Participants usually felt that their maps were complete in about 10 min and that their SVGs had reached their end in about an hour. However, as more time was devoted to SVGs, more data could be collected. Ultimately, we suggest that the central reason for SVGs resulting in the identification of more spatially specific location and with much greater depth is because of the sensory stimulation. It triggers not only more spaces, but also places that facilitate or mitigate exposures (such as a check cashing store or rehabilitation center), and it also reveals spatial interconnections and processes over space and time. In other words, it is a more evolved geographic context. For example, we can use the SVG to help explain why areas B, D and F in Figure 2 were not identified through the sketch map or calls for service data. In the case of the sketch maps, these areas were literally outside the map provided to the participants. In almost all the SVGs, the police are described as not to be trusted and that there is a code on the street for not reporting events, especially in the highest crime areas of the neighborhood. There are also streets identified that the participants claim are police no-go areas. As a result, the 911 calls for service data are likely to contain geographic data gaps where only the most severe situation generates a data point. The ex-offender SVG not only gives us more contextual data, but also can fill in some of these data gaps left by official sources.

We will now consider the conveyed structure of the SVG before delving deeper into how environmental stimuli affect the contextual map.[7] From a purely space-time per-spective, the neighborhood is no longer seen as one space but rather as nested, with varying degrees of detail based on what is seen and the past experiences of the participant in specific places. While there may be general consensus among the parti-cipants about area A in Figure 2, the experiences within and therefore the specific locations identified vary. Interestingly, the SVG 'cruise' also mimics the typical activity for many within the neighborhood – the participant knows where to go generally and then lets various cues refine the search. For this study, it included going to a known area (A) and then describing houses/bars/corners, etc. For a drug seeker, it would also include going to a known street and then refining the search based on visual cues. It should also be noted that the space within the various layers of the nest is also complex, especially in terms of heterogeneity. A road may contain both 'bad' and 'good' exposures, some-times being located next to each other.[8] This level of fine scale heterogeneity is challenging to replicate on a sketch map.

A contributing factor within these nests is spatial interconnections, such as the alley behind, the proximate public housing and the lack of a fence, all of which make that location problematic, and therefore to reduce the location to a single point would be missing the importance of spatial setting. Similarly, that geographic setting also has a temporal dimension: the changing nature of a park at night, the danger around a corner store during the evening hours or the changing activity along a road across the course of a day. There are even changes with seasons, or more importantly as temperatures increase, so the activity around corner stores, bars and even on front porches increases and can lead to more violence. This appreciation of time largely goes unacknowledged on the sketch map unless specifically requested. There is also a personal time continuum where experiences of one location will shift fluidly between the now, recent and distant pasts. From a GIScience perspective, this leads to the question of how to identify and then record such temporal mixing in the attribute table. This can include temporal uncertainty in terms of the participant being away (in prison) and therefore not being sure whether the described location still holds the same relevance.[9] Even more challenging is how a single space can also contain multiple memories; a street is described as being happy (maybe from childhood), but specific events (murder) and locations (drug houses) caused a decline, and then more recently, the situation has improved, especially in comparison with other neighboring streets.[10]

What is core to all these descriptions, and what ultimately differentiates the SVG from the sketch map, is the environmental cue. While the criminal experience of the participants varied, a connecting factor was drug addiction. Where to find drugs, how to acquire money to purchase drugs (eg through drug selling, burglary or prostitution) and the associated acts of violence, despair and loss (especially with discussions about the impact on children), all of which were described in the SVG. Most of the stimuli on the SVG ride acted as a window into each of these. Common perception and personal experience are also more easily separated in the SVG. While participants may identify spaces on a sketch map through what they have heard or believe, the SVG is far less forgiving in terms of allowing such general observations.

There is considerable variation in terms of what is an environmental cue. The simplest cue might be a house, but the attributes observed and interpreted by the participant lead to a deepening of the description through experience of that location or street experience in general: the trash in the yard, the way two or three blinds in the window are bent back, which is a likely indication that the house is being used for drugs. These visual cues are vital in developing the context of a scene that emerges through the SVG, but cannot possibly be captured on the sketch map, at least not in any detail. For example, one participant had identified the street being driven as problematic, but seeing bullet holes deepened the story at that location, while children playing nearby extended the participant's comment, bringing in other ecological factors such as time of year.

> "… cause that corner building up here sells heroin, right here, and something happened right there at the house across the street, that white one it has bullet holes in the side window… this one that has the board on it, they used to live there, but that guy sells heroin, isn't that crazy…look at X house right here has bullet, it has little bullet holes see them. I don't think (for children this area is) safe, because anything could happen, cars could race down the street especially when the drug, gang activity starts getting bad in the summer."

This previous quote contains two further observations worthy of discussion, people as visual cues and empathy. Not only will the participant interpret the built environment but activity within (cars outside a house, the way two cars are parked in a lot, a woman walking down the street) will also lead to a spatially specific mention. Sometimes it is a person being in proximity to a known building, other times it is the reverse, a known individual (usually a sex worker) interacting with a house, car or lot. Some of these human cues are known to the participant, while others display the trappings and activities that are well *known on the street*. Interestingly, several potential crimes, such as an interaction between a sex worker and a client, or a possible drug exchange, were witnessed while conducting the SVG by the participant that would have gone unnoticed by the research team.[11]

The second observation concerns empathy. In the above quote, the participant links visual cues, with past experiences, and the presence of children playing to create a deeper context on that space. In another example, the participant was asked about a pocket park, which led to the response that it was 'safe' during the day but had heavy drug use at night. The participant went on to mention *why would any child be in there at night*, then paused and reflected, that of course there was the possibility of children finding needles the next day. Not only does the SVG reveal a deeper insight, but the participant can even make these types of connections, from his/her perspective *because* of being involved in the SVG collection.

The investigation of the SVGs has revealed several other topics requiring further investigation outside the scope of this paper. It would be interesting to look at inter-participant variation in the amount of detail described, with some commenting on every house, and even details within a house, while others were more general in their description of a space. The participants' appreciation of spatial relations also varied, for example, some would think of boundaries (usually roads), and how crimes vary spatially. It would be interesting to see how much of these variations are due to the experiences of the participant. For example, one driver for a gang/drug dealer would cover larger areas of the city and so his knowledge was more hierarchical and nested, whereas the single neighborhood drug user/sex worker had a more detailed under-standing of smaller areas. That second participant would interpret micro places through the lens of a sex worker's state of mind, how she would take a client to a bush or an abandoned house, purely on the promise of drugs and/or the likelihood of not being arrested. By her own admission, the danger of these spaces did not enter her thinking. A third participant, also addicted to drugs, would use theft to fund her habit, and would stay away from abandoned houses because of the danger they posed. The SVG approach offers an interesting opportunity to begin a new line of research into how perceptual mapping variations are influenced by these types of experience.

Further research should also consider issues of spatial precision. While this has already been discussed with regard to sketch maps in terms of how different markings capture the underlying geography (Curtis *et al.* 2014), more work is needed on spatial specificity in the SVG. For example,

like I don't know, rape and kill. They stabbed that girl in that field right over here behind that X Street express, a whole bunch of times

see this X Street Express down here? There was a field behind it and that's where they
stabbed that girl a bunch of times

Using this example, we have potential issues with where the comment should be attached
to, where the Wordmapper interpolation procedure actually places the comment (a function of both the model and the route being on a road), the precision of the GPS unit and, is
often the case, a distance across which the comment occurs. Similarly, the size of the area (a
pocket park compared to a building) or the importance of the extended spatial setting (the
contributing environment around the house) makes us question the relevance of using just
a single point. To counter this, for this paper we used three buffer sizes around each spatially
specific location to approximate this fuzziness. While it is possible to manually shift each
comment to the exact location being described because of the simultaneously collected
video, it would be more effective to develop a series of placement guidelines or even rules
to facilitate a more automated mapping of SVG context.

In this paper, we only focus on spatially specific mentions in the SVG, where a sensory
cue leads to context. In so doing, we miss other spatial insights contained in the fuzzy or
spatially inspired comments. Examples include comparisons among areas, transition
zones or boundaries based on activities that occur in these locations, or where there are
concentrations of activity, such as a road known for prostitution, or where homeless
camps can be found. For example, in this spatially inspired comment, there is also specific
spatial detail. We should consider how we can incorporate these additional spatial data.

" you're in that neighborhood because there's drugs in that neighborhood and you're
walking down that street probably to get drugs on your way to get drugs...I mean you
can be picked up on any street at any time, certain neighborhoods, guys will just.. you
know, I get real uncomfortable on A street, there's a certain part of it, every time I walk on it,
they don't just stop with no.."

While we mentioned the presence of emotion on the ride, and even empathy shown by the
participant, we have not performed any sentiment analysis on these SVGs, and yet, arguably,
these data are ideal for this type of research. Not only do we have a whole narrative to train
our interpretation of what is transcribed, but we can also return to the audio to hear how the
words are spoken. For example, one subject commented on the recall beginning to make
her cry, whereas another suddenly felt uncomfortable and wanted to leave, this desire for
flight triggered by the stories he was telling. These data open research avenues into how
such emotions can be effectively transcribed and mapped, as well as into how to recognize
linguistic constructs that are indicative of an emotional type, including sarcasm.

In terms of what these contextual insights mean for crime analysis, the mixing of
various crime types within the same space suggests that there is a danger in focusing on
a single crime type – these are multi-risk environments and should be treated as such.
As an illustration of this, although drugs are mentioned on all but one of the sketch
maps, there is no indication from the way the participants mark the maps that drugs are
an integral force in this neighborhood – they appear consistently, but so do other crimes
listed (eg prostitution, various forms of assault). The SVGs make clear that the crimes and
conditions in the built and social environment are driven by drugs. This result indicates
the value of being embedded in place with participants so that the environment can cue
their insights.

Notes

1. See Kwan and Ding (2008) for a thorough literature review on narrative analysis.

2. While gaining the perspective and insights of those who best understand an area or topical situation has many benefits, it should also be noted that there are dangers that must be considered with regard to the welfare of the participant and those being discussed. The SVG can cause emotion in the subject, and while the team discussed both before and during the ride that the participant could end the SVG at any point, we must also be mindful of causing any emotional trauma. This will vary in every situation, but during the rides described in this paper, no participant ever described anything other than an unease – not because of an immediate threat but because of what a space had meant to that person in the past. Even so, embeddedness is not always desirable or even possible, which raises the need for investigation of Virtual Reality (VR) or other forms of simulation as an alternative approach in this line of research. We must also be aware of any stigma that can be attached to a person, building or neighborhood in general through this work. To counter this, no individuals have been mentioned, nor any places identified on a map. Even the neighborhood being described and mapped has been removed. As use of geographically embedded approaches grows, inquiry should also expand to focus on the numerous ethical considerations that arise in order to provide guidance to protect our participants while making the best use of their valuable local knowledge.

3. Wordmapper, developed by A. Curtis and Ajayakumar at Kent State University, was designed to provide mapping capability of the transcribed text for experienced spatial users (GIS output), spatial novices (Google Earth output) and researchers wanting to perform qualitative analysis (Comma Separated Value (CSV) output). Wordmapper, a Python-based stand-alone software, has five modules including preprocessing, combination, visualization, query and output. The preprocessing module accepts the narrative in the form of transcribed text, GPS data in the form of Comma Separated Value (CSV) file and offset time to synchronize between the temporal information in the narrative and GPS data. The preprocessing module also validates all data inputs. The combiner module syncs the narrative with the GPS data based on timestamps to create tuples of narrative texts of the form <sentence,timestamp,coordinates>. Apart from mapping the narrative and the GPS data, the combiner module also extracts words from sentences and assigns coordinates to each of them. The words are extracted using the tokenizer module from the Natural Language Toolkit (NLTK) module in Python. Each word of the form <text, time> is converted to a triplet of the from <text,time,coordinates> using a linear interpolation algorithm. The visualization module in Wordmapper receives processed geonarratives from the combiner module and utilizes Google Maps API and keyword-based wordclouds to visualize the data. Output from Wordmapper includes interactive mapping of words, comments and keyword queries, either in the onscreen map or as KML or shapefile outputs.

4. Even though KDE is a useful technique for visualization, in typical application it is not a statistically rigorous analytical tool (Curtis *et al.* 2010). To improve statistical rigor, the Spatial Filter module utilizes a variant of spatial filtering developed by Rushton and Lolonis and mainly used in epidemiological studies (Rushton and Lolonis 1998, Rushton *et al.* 2004). A grid is imposed over the study area, and a filter (circle) with an input adjustable bandwidth is placed on every grid node. At each grid node, rates are calculated using numerator (eg wordmapper keywords) and denominator (total words). The locations of numerator and denominator points create a smooth surface rather than one truncated by political boundaries. With a larger bandwidth, more points are included in the rate calculation, and the final surface tends to be smoother. Spatial filter output consists of a grid with a rate calculation attached to each node. A Monte Carlo simulation is used as a proxy for a statistical significance test. Synthetic numerator sets are created from the denominator sets, and rates are calculated at each node for each simulated surface. Based on the number of simulations, typically 99,999, and 9999, the *p*-values can be

calculated for each grid based on the following formula (Supplementary Material – Appendix B). As the method can be computationally expensive based on the granularity of the grid and the extent of the study area, we have developed a stand-alone spatial filter application using Python, which utilizes a spatial k-d tree for neighbor-based searches and eliminates grids without neighbors for simulations.

5. Note that in small areas where many spatially specific comments were made, the individual buffers of such comments are merged.

6. Individual sketch map locations of high crime, along with SVG paths and locations of high crime, are provided in Supplementary Material – Appendix C.

7. We have purposefully stayed away from crime theory in this paper. However, the detail in the SVG also has value in terms of supporting various theories and debates in the criminology literature. For example, the narratives are full of 'broken windows'-style descriptions, usually in terms of explaining why blighted and rundown properties act as an attractor for crime, but also as a precursor to other activity: a visual cue that the street is sliding downward. Connected to this, the SVG frequently mentions the importance of neighborhood oversight, which is a key component of routine activity theory.

'I think that abandoned houses are pretty scary, the more you have in a neighborhood the more activity you are going to... It's about the neighborhood and who lives by it, you got a house like this where they take care of their yard and they got a fence, a lot of times they are going to pay attention and they are going to call the cops so somebody might avoid that, so that you know, the cops just come in there and arrest them.'

Interestingly, this subject then went on to describe how visible presence shouldn't fool you as many drug dealers live in the nicer homes so as not to draw attention to themselves (or even in completely different and better neighborhoods) and deal from cars or 'trap houses'.

8. Again, the value of the SVG is that it can offer explanations to this heterogeneity. For example, locating a drug operation near a sanctuary house is attractive to some dealers as they know there is a desperate clientele trying hard not to relapse. *'..this is the (a sanctuary house), there is a dope house, and right there on the left. There is a dope house right here on the left, two doors down.'*

9. While not described in this paper, the SVGs contain multiple references to the fluidity of crime in the neighborhood, in terms of how people seek (eg drugs), how drug houses change, how drug organizations reorder themselves spatially and the difference between the home and activity space (for professional drug dealers).

10. All our participants displayed examples of this movement between the present and the past. It would be interesting to investigate how much of this mixing also influences the sketch maps.

11. This is a difference with the seminal work of Sampson and Raudenbush (1999) who had commented that it was unlikely to ever witness an actual criminal event happening. These SVG reveal that many such events occur on every ride if there is an expert who can interpret the environment.

Acknowledgments

The opinions, findings and conclusions or recommendations expressed in this paper are those of the authors and do not necessarily reflect those of the Department of Justice.

In addition, we appreciate the contributions of Lauren Porter and the constructive comments provided by the reviewers.

Disclosure statement

No potential conflict of interest was reported by the authors.

Funding

This work was supported by the National Institute of Justice [2013-R2-CX-0004], awarded by the National Institute of Justice, Office of Justice Programs, U.S. Department of Justice.

ORCID

Jacqueline W. Curtis ⓘ http://orcid.org/0000-0001-6046-6476
Jayakrishnan Ajayakumar ⓘ http://orcid.org/0000-0001-9564-7728
Susanne Mitchell ⓘ http://orcid.org/0000-0003-1052-4345

References

Anderson, J., 2004. Talking whilst walking: a geographical archaeology of knowledge. *Area*, 36, 254e261. doi:10.1111/j.0004-0894.2004.00222.x

Anderson, J. and Jones, K., 2009. The difference that place makes to methodology: uncovering the 'lived space'of young people's spatial practices. *Children's Geographies*, 7 (3), 291–303. doi:10.1080/14733280903024456

Bell, S.L., *et al.*, 2015. Using GPS and geo-narratives: a methodological approach for understanding and situating everyday green space encounters. *Area*, 47 (1), 88–96. doi:10.1111/area.2015.47. issue-1

Bell, S.L., Wheeler, B.W., and Phoenix, C., 2017. Using geonarratives to explore the diverse temporalities of therapeutic landscapes: perspectives from "Green" and "Blue" settings. *Annals of the American Association of Geographers*, 107 (1), 93–108. doi:10.1080/24694452.2016.1218269

Bergeron, J., Paquette, S., and Poullaouec-Gonidec, P., 2014. Uncovering landscape values and micro-geographies of meanings with the go-along method. *Landscape and Urban Planning*, 122, 108–121. doi:10.1016/j.landurbplan.2013.11.009

Beyer, K.M., Comstock, S., and Seagren, R., 2010. Disease maps as context for community mapping: a methodological approach for linking confidential health information with local geographical knowledge for community health research. *Journal of Community Health*, 35 (6), 635–644. doi:10.1007/s10900-010-9254-5

Boschmann, E.E. and Cubbon, E., 2014. Sketch maps and qualitative GIS: using cartographies of individual spatial narratives in geographic research. *The Professional Geographer*, 66 (2), 236–248. doi:10.1080/00330124.2013.781490

Butler, M. and Derrett, S., 2014. The walking interview: an ethnographic approach to understanding disability. *Internet Journal of Allied Health Sciences and Practice*, 12 (3), 6.

Carpiano, R.M., 2009. Come take a walk with me: the "go-along" interview as a novel method for studying the implications of place for health and well-being. *Health & Place*, 15, 263e272. doi:10.1016/j.healthplace.2008.05.003

Curtis, A., *et al.*, 2015. Spatial video geonarratives and health: case studies in post-disaster recovery, crime, mosquito control and tuberculosis in the homeless. *International Journal of Health Geographics*, 14 (1), 22. doi:10.1186/s12942-015-0010-z

Curtis, A., *et al.*, 2016. Context and spatial nuance inside a neighborhood's drug hotspot: implications for the crime–health nexus. *Annals of the American Association of Geographers*, 106 (4), 819–836. doi:10.1080/24694452.2016.1164582

Curtis, A., Duval-Diop, D., and Novak, J., 2010. Identifying spatial patterns of recovery and abandonment in the post-Katrina Holy Cross neighborhood of New Orleans. *Cartography and Geographic Information Science*, 37 (1), 45–56. doi:10.1559/152304010790588043

Curtis, J.W., *et al.*, 2014. The prospects and problems of integrating sketch maps with geographic information systems to understand environmental perception: A case study of mapping youth fear in Los Angeles gang neighborhoods. *Environment and Planning B: Planning and Design*, 41 (2), 251–271. doi:10.1068/b38151

Curtis, J.W., 2016. Transcribing from the mind to the map: tracing the evolution of a concept. *Geographical Review*, 106 (3), 338–359. doi:10.1111/gere.2016.106.issue-3

Doran, B.J. and Burgess, M.B., 2011. *Putting fear of crime on the map: investigating perceptions of crime using geographic information systems*. New York: Springer Science & Business Media.

Doran, B.J. and Lees, B.G., 2005. Investigating the spatiotemporal links between disorder, crime, and the fear of crime. *The Professional Geographer*, 57 (1), 1–12.

Downs, R.M. and Stea, D., eds., 1974. *Image and environment: cognitive mapping and spatial behavior*. New Brunswick: Transaction Publishers.

Elwood, S. and Martin, D., 2000. "Placing" interviews: location and scales of power in qualitative research. *The Professional Geographer*, 52, 649e657. doi:10.1111/0033-0124.00253

Evans, J. and Jones, P., 2011. The walking interview: methodology, mobility and place. *Applied Geography*, 31 (2), 849–858. doi:10.1016/j.apgeog.2010.09.005

Gould, P. and White, R 1974. *Mental maps*. London: Penguin.

Gould, P.R. and White, R.R., 1968. The mental maps of british school leavers. *Regional Studies*, 2 (2), 161–182. doi:10.1080/09595236800185171

Jones, P., *et al.*, 2008. Exploring space and place with walking interviews. *Journal of Research Practice*, 4, np.

Krystosik, A.R., *et al.*, 2017. Community context and sub-neighborhood scale detail to explain dengue, chikungunya and Zika patterns in Cali, Colombia. *PloS One*, 12 (8), e0181208. doi:10.1371/journal.pone.0181208

Kwan, M.P., 2012a. The uncertain geographic context problem. *Annals of the Association of American Geographers*, 102 (5), 958–968. doi:10.1080/00045608.2012.687349

Kwan, M.P., 2012b. How GIS can help address the uncertain geographic context problem in social science research. *Annals of GIS*, 18 (4), 245–255. doi:10.1080/19475683.2012.727867

Kwan, M.P., 2013. Beyond space (as we knew it): toward temporally integrated geographies of segregation, health, and accessibility: space–time integration in geography and GIScience. *Annals of the Association of American Geographers*, 103 (5), 1078–1086. doi:10.1080/00045608.2013.792177

Kwan, M.P. and Ding, G., 2008. Geo-narrative: extending geographic information systems for narrative analysis in qualitative and mixed-method research. *The Professional Geographer*, 60 (4), 443–465. doi:10.1080/00330120802211752

Kyttä, M., *et al.*, 2013. Towards contextually sensitive urban densification: location-based softGIS knowledge revealing perceived residential environmental quality. *Landscape and Urban Planning*, 113, 30–46. doi:10.1016/j.landurbplan.2013.01.008

Lager, D., Van Hoven, B., and Huigen, P.P., 2013. Dealing with change in old age: negotiating working-class belonging in a neighbourhood in the process of urban renewal in the Netherlands. *Geoforum; Journal of Physical, Human, and Regional Geosciences*, 50, 54–61. doi:10.1016/j.geoforum.2013.07.012

López, N. and Lukinbeal, C., 2010. Comparing police and residents' perceptions of crime in a phoenix neighborhood using mental maps in GIS. *Yearbook of the Association of Pacific Coast Geographers*, 72 (1), 33–55. doi:10.1353/pcg.2010.0013

Lynch, K., 1960. *The image of the city*. Cambridge: MIT Press.

Manton, R., *et al.*, 2016. Using mental mapping to unpack perceived cycling risk. *Accident Analysis & Prevention*, 88, 138–149. doi:10.1016/j.aap.2015.12.017

Mennis, J., Mason, M.J., and Cao, Y., 2013. Qualitative GIS and the visualization of narrative activity space data. *International Journal of Geographical Information Science*, 27 (2), 267–291. doi:10.1080/13658816.2012.678362

Merriman, P., 2014. Rethinking mobile methods. *Mobilities*, 9 (2), 167–187. doi:10.1080/17450101.2013.784540

O'Neill, E., *et al.*, 2015. Exploring a spatial statistical approach to quantify flood risk perception using cognitive maps. *Natural Hazards*, 76 (3), 1573–1601. doi:10.1007/s11069-014-1559-8

Rushton, G., *et al.*, 2004. Analyzing geographic patterns of disease incidence: rates of late-stage colorectal cancer in Iowa. *Journal of Medical Systems*, 28, 223–236.

Rushton, G. and Lolonis, P., 1998. Exploratory spatial analysis of birth defect rates in an urban population. *Statistics in Medicine*, 15, 717–726. doi:10.1002/(SICI)1097-0258(19960415)15:7/9<717::AID-SIM243>3.0.CO;2-0

Sampson, R.J. and Raudenbush, S.W., 1999. Systematic social observation of public spaces: a new look at disorder in urban neighborhoods. *American Journal of Sociology*, 105 (3), 603–651. doi:10.1086/210356

Schuch, L., Curtis, A., and Davidson, J., 2017. Reducing lead exposure risk to vulnerable populations: a proactive geographic solution. *Annals of the American Association of Geographers*, 107 (3), 606–624. doi:10.1080/24694452.2016.1261689

Warren, S., 2017. Pluralising the walking interview: researching (im) mobilities with Muslim women. *Social & Cultural Geography*, 18 (6), 786–807. doi:10.1080/14649365.2016.1228113

Travel impedance agreement among online road network data providers

Eric M. Delmelle ⓘ, Derek M. Marsh, C. Dony and Paul L. Delamater ⓘ

ABSTRACT

Online mapping providers offer unprecedented access to spatial data and analytical tools; however, the number of analytical queries that can be requested is usually limited. As such, Volunteered Geographic Information (VGI) services offer a viable alternative, provided that the quality of the underlying spatialtheir data is adequate. In this paper, we evaluate the agreement in travel impedance between estimates from MapQuest Open, which embraces OpenStreetMap (OSM) data–a is based on VGI datasetfrom OpenStreetMap (OSM), and estimates from two other popular commercial providers, namely Google Maps™ and ArcGIS™ Online. Our framework is articulated around three components, which simulates potentialcalculates shortest routes, estimates their travel impedance using a routing service Application Program Interface (API), and extracts the average number of contributors for each route. We develop an experimental setup with a simulated dataset for the state of North Carolina. Our results suggest a strong correlation of travel impedance among all three road network providers. and that travel impedanceThe agreement is the greatest in areas with a denser road network and the smallest for routes of shorter distances. Most importantly, tTravel estimates from MapQuest Open are nearly identical to both commercial providers when the average number of OSM contributors along the route is larger. The latter finding contributes to a growing body of literature on Linus's law, recognizing that a larger group of contributors holds the potential to validate and correct inherent errors to the source dataset.

1. Introduction

The concept of spatial proximity has received considerable attention in GIScience and various domains, such as public health (Casas *et al.* 2017, Kirby *et al.* 2017), food access (Widener *et al.* 2011; Racine *et al.* 2018), emergency management (Peleg and Pliskin 2004; Murray and Tong 2009), urban and transportation planning (O'Sullivan and Morrall 1996; Dony *et al.* 2015) and location modeling (Murray 2010) to cite a few. Core to this concept is the notion of distance, which represents the space separating two objects with distinct coordinates. Although straight-line distances have traditionally been used due to the ease of its calculation (Phibbs and Luft 1995; Boscoe *et al.* 2012), network-based travel measures are more representative as they incorporate topological structures (Apparicio *et al.* 2008; Shahid *et al.* 2009; Jones *et al.* 2010; Delmelle *et al.* 2013). However, their estimation is generally more time-consuming since a GIS-based network must be prepared (Gutiérrez and García-Palomares 2008; Delmelle *et al.* 2013).

Travel impedance is generally estimated by minimizing the cost of travel from an origin to a destination through a series of nodes and edges. This minimization approach is referred to as the shortest path problem and can be solved using the Dijkstra (1959) algorithm. Travel impedance is typically reported as travel distance, defined by the sum of the length of each traveled edge, or travel time, defined by the sum of the length of each traveled edge divided by its maximum allowed travel speed (Wang and Xu 2011; Delamater *et al.* 2012). Travel estimates can be inaccurate due to a myriad of reasons, such as incomplete or outdated network data, or geocoding uncertainty (Berke and Shi 2009; Delmelle *et al.* 2013). The impact of the latter is illustrated in Figure 1, where such errors can be introduced during the snapping process on a road network. An imprecise point coordinate and/or a large snapping tolerance could geocode the point at one of several locations depending on how many road segments fall within the probable area, with the risk of influencing the resulting travel estimates.

The shortest path problem forms the basis of routing algorithms used by online network data providers. While routing functionality is available through Desktop GIS software (e.g. ArcMap, ESRI, Redlands, CA), defining an accurate road network can

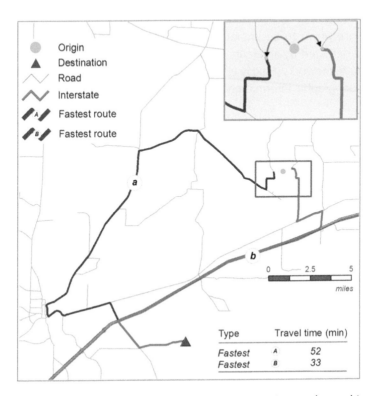

Figure 1. Snapping of a geocoded point. Snapping parameters can impact the resulting travel time, here for a hypothetical origin-destination.

require extensive pre-processing efforts such as incorporating speed limits, honor one-way restrictions, turn penalties, reflect connectivity among roads and testing its accuracy (Delmelle *et al.* 2013). As the study area grows larger, so does the computer's storage to host the road network and its underlying data; and the higher the computer's performance will need to be to efficiently calculate thousands of shortest paths (i.e. run optimization algorithms). For a road network covering a large area and for a project that requires calculating thousands of routes, a commercial Desktop GIS software could potentially crash. Online network data providers like Google Maps™ and Bing Maps™ can eliminate this time-consuming pre-processing task because their network is up-to-date, as speed limits, restrictions and connectivity are already incorporated in their routing service Application Program Interface (API) (Wang and Xu 2011; Boscoe *et al.* 2012). Users can use a programming language (such as Python) to connect to the server of the provider which hosts their road network data. Users can make 'route requests'; the server sends back a response, which includes the path with the least impedance. Since calculations happen on a remote server, the user's computer capacity and performance do not matter. However, these proprietary online road network data providers generally limit the number of analytical requests an unlicensed user can make, which make them impractical for sizeable spatial analysis.

Alternative to proprietary providers are those based on Volunteered Geographic Information (VGI), given that the quality of the underlying spatial data is adequate for the research intent. VGI refers to the phenomenon by which users, most with no formal training, contribute geospatial data (Goodchild 2007; Corcoran *et al.* 2013). User-generated content is a form of crowd-sourcing where information is acquired that would otherwise prove to be impractical to obtain by other means. Data collection is voluntary and gathered by amateurs without central coordination or strict adherence to particular data frameworks (Haklay *et al.* 2010). This practice has proven to be successful for acquiring geographic information of different scales in a timely manner and at a very low cost.

One of the most utilized, analyzed, and cited VGI platforms for road data, with an increasing popularity over the past few years, is OpenStreetMap (OSM) (Neis and Zielstra 2014; Arsanjani *et al.* 2015). Its success can be attributed to a strong and growing community support along with a rapidly increasing demand for non-proprietary geospatial data (Jilani *et al.* 2013a, 2013b). OSM has received considerable attention in the last decade as it provides an alternative to commercial and authoritative data (Arsanjani *et al.* 2015). OSM datasets have been used in various domains such as disaster relief in Haiti (Zook *et al.* 2010; Soden and Palen 2014), fine resolution population estimations (Bakillah *et al.* 2014), updating of Digital Elevation Models combining up-to-date OSM data with high-resolution provided by the Airborne Laser Scanning (Klonner *et al.* 2015), and routing (Schmitz *et al.* 2008; Goetz and Zipf 2012; Neis 2015).

1.1. *Data uncertainty in OSM*

Despite its growing popularity, OSM data brings an important challenge with regards to data uncertainty (Ciepluch *et al.* 2010; Mooney *et al.* 2010; Jilani *et al.* 2013a), which has long been a core principle of GIScience (Zhang and Goodchild 2002; Griffith 2018). There are two properties of data uncertainty: accuracy and precision. Accuracy is the degree to which a measured value approaches the true value, while precision reflects the dispersion of measured values around the mean value for that group of measurements (GPS precision is subject to the accuracy of the measuring device). *Positional accuracy* refers to the spatial components of an object, where a higher accuracy suggests that an object is located closer to its true position on Earth. In a comparison of OSM to the Ordnance Survey of Great Britain, Haklay (2010) found an average of 6-m point displacement within study sites across London. Similar levels of positional accuracy was corroborated by Girres and Touya (2010) in their comparison of the French OSM dataset to the French National Mapping Agency (Institut Géographique National, IGN) BD Topo dataset; point positional displacement was on average 6.65 m and the average distance between the two roads networks was 2.19 m.

The quality of the OSM road network is a significant determinant for its applicability as a routing and navigation application, and recent literature has suggested that current OSM coverage may not be sufficient (Ciepłuch *et al.* 2010). *Completeness* is an indication of dataset comprehensiveness, where a complete dataset includes all real-world elements and all of the objects' pertinent information, or attributes (Hochmair *et al.* 2015). Proprietary geospatial datasets, such as Google Maps[1] and Bing Maps, hold a distinct advantage in regards to completeness (Ciepluch *et al.* 2010). The standardized and centralized data collection of proprietary online network data providers offer consistency and some assurance of a completed dataset, however they lack transparency concerning data acquisition and maintenance. To alleviate some areas of low user contribution and poor coverage in OSM, publicly accessible datasets from authoritative sources have been used, such as TIGER data in the USA[2] However, bulk imports of authoritative data may hinder contribution efforts in remote or rural regions, where imported datasets are less scrutinized and improvement of the road network is less likely to occur.[3]

A high number of active OSM contributors leads to a quality OSM dataset, that is likely to be more *temporally accurate* (Girres and Touya 2010; Haklay *et al.* 2010). In areas where OSM has a high contributor count, local knowledge may be more effective in mapping new points of interest or removing outdated ones. Therefore, OSM may have an advantage over proprietary geospatial data in terms of temporal accuracy (local, significant, and/or new features may be missing from authoritative or proprietary datasets because local knowledge provides data that would not be known as quickly outside the area, see for instance Zook *et al.* (2010) in Haiti). Yet, the assumption that OSM is of lower quality than proprietary maps can undermine the trust in new information that is present in OSM (Jilani *et al.* 2013b; Jilani *et al.* 2013b).

1.2. *Linus' law*

Since participants collecting OSM data potentially lack any formal training in geographic data collection, central coordination is weak to non-existent. Adherence to a particular data structure is not required, no assumptions can be made about the overall quality of the data (Goodchild and Li 2012). As Zhang and Goodchild (2002) point out, 'uncertainty exists in the whole process from geographical abstraction, data acquisition, and geoprocessing to the use of data'. Thus, errors in the data are likely to propagate and further contaminate subsequent geospatial analysis.

Nevertheless, and even with known precision limits of the measurement device (e.g. digits after the comma on GPS devices), repeated observations/measurements can increase data accuracy (central limit theorem). As such, we hypothesize that a greater number of OSM contributors can lower data uncertainty. The idea that a group of contributors contain the ability to validate and correct the errors that an individual might make to converge on the truth is known as *Linus' Law* (Haklay *et al.* 2010). For the OSM dataset, 'if one individual contributes an error, others can be expected to edit and correct the error, and the success of this mechanism rises in proportion to the number who look at the contribution' (Goodchild and Li 2012).

The application of *Linus' Law* to VGI is not new and when evaluating the OSM dataset, the literature has rendered largely positive results, especially in terms of positional accuracy (Keßler and de Groot 2013). Haklay *et al.* (2010) have suggested that the positional accuracy improved with an increase in the number of contributors up to a threshold ($n > 13$) at which improvement stabilized. Girres and Touya (2010) confirmed *Linus' Law* in terms of dataset completeness: the number of OSM objects in an area grows nonlinearly in relation to the number of contributors in the area.[4]

1.3. *Research questions*

Although there exist several studies which have demonstrated that a greater number of OSM contributors will likely increase the spatial agreement between OSM and reference datasets, little is known how these contributors can impact the accuracy of travel estimates. Further, very few papers have compared the similarities of travel estimates between different online network data providers (Delmelle *et al.* 2013; Socharoentum and Karimi 2014). It is in this context of data uncertainty that our paper is situated, proposing a methodological framework to compare data from proprietary routing providers (Google Maps™ and ArcGIS™ Online) with a VGI routing provider (MapQuest Open). In the light of these objectives, our research attempts to answer the following questions:

- What is the degree of agreement in travel impedance estimates among online road network data providers, and what are the implications of such uncertainty for travel estimates?
- Do routes calculated using OSM data present significantly different travel impedance estimates in comparison to proprietary online spatial datasets?
- Does a correlation exist between the number of contributors and the aforementioned degree of uncertainty, and how does it relate to the *Linus' Law*?

Section 2 provides a description of the methodological framework to determine the agreement between proprietary and VGI routing service providers. This method extracts travel impedance between origin-destination (OD) pairs using the API of three online routing services. The difference in travel impedance is measured for a large number of OD pairs, simulated over different ranges of distances in the state of North Carolina, USA Results are discussed in Section 3, followed by a discussion and conclusion, in Section 4. The findings of this research are particularly important in evaluating the (1) reliability of OSM for routing applications and (2) whether a larger number of OSM contributors can lead to travel estimates similar to the ones obtained using a complete and up-to-date provider such as Google Maps.

2. Methodology and experimental setup

The methodology section is articulated around (1) online road network routing web services, (2) VGI user contribution and (3) assessment of network travel impedance and associated uncertainty. We use simulated pairs of origins and destinations across the state of North Carolina.

2.1. *Online network routing web services*

To estimate travel impedance uncertainty using the OSM road network, two proprietary online network providers, namely Google Maps and ArcGIS Online,[5] are selected for comparison purposes. Although OSM does not provide routing functionality, MapQuest Open provides a routing service comparable to the two reference datasets utilizing the OSM dataset.[6] The flowchart of the methodology is provided in Figure 2.

2.1.1. *Application programming interface (API)*
Online network data providers can deliver services such as geocoding and routing through an API. Focusing on routing services, requests are constructed as a Uniform Resource Locator (URL) string, which includes the network data provider server web address, the origin and destination, and potentially other calculation specifications. A routing service API allows a user to process batches of data through these services in a timely, simplified fashion, sending and receiving data via Hypertext Transfer Protocol (HTTP).

2.1.2. *Implementation in python*
The Python programming language is used to employ this API 'automation' technique for large numbers of routing calculation requests. An origin-destination (OD) matrix provides begin and end locations for each route (Wang and Xu 2011). For this analysis, all routing requests disregard traffic data. In doing so, travel impedance estimates only reflect the providers' network and are not influenced by congestion, construction, or

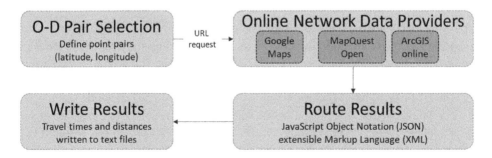

Figure 2. Flowchart depicting the steps required to estimate travel impedance using an OD matrix and online network data providers.

other temporal factors. Most specifications were kept as default including routing optimization, set to calculate a route, which results in the shortest possible travel time. A URL request is sent to each corresponding online network data provider via Python's urllib2 built-in library.

2.1.3. Outputs

Routing results are returned in either JavaScript Object Notation (JSON) or Extensible Markup Language (XML). Both formats allow rapid and efficient parsing within Python, allowing all extemporaneous routing data to be ignored and travel impedance values to be extracted and then stored. For each routing calculation, travel distance and travel time are extracted. Final results include the OD index, the origin and destination coordinates, and the travel distance and travel time for each online network data provider stored as a text file for further analysis.

2.2. Assessments of network travel impedance

The correlation between the travel distances and times extracted from each respective online network data providers is calculated using the R programming language. Because the travel impedance estimates are independent of one another, the correlation coefficient r value is used for all uncertainty assessments; however, coefficient of determination r^2 is provided for reference purposes. The slope of the correlation is used to indicate under- or overestimation between providers. Finally, differences in travel distance estimation are calculated as a percent and used for comparative analysis.

$$\Delta d_{ij} = \frac{|d_{ij,k=1} - d_{ij,k=2}|}{(d_{ij,k=1} + d_{ij,k=2})/2} \tag{2}$$

With i the origin location, j the destination location and k the online network data provider. The term $d_{ij,k=1}$ is the calculated network distance from i to j using network provider k; for $k = \{1, 2, 3\}$. This calculation is applied for all iterations of k, resulting in three individual percent measures of combination $\{k = 1, k = 2\}$; $\{k = 2, k = 3\}$; $\{k = 1, k = 3\}$.

2.3. Generating an origin-destination matrix

Origin and destination points are carefully selected with two objectives: (1) uniform representation of the range of possible distances–dictated by the extent and geometry of the study area, and (2) spatial coverage of the study area. The selection of candidate origin and destination points is completed using ArcGIS Desktop (ESRI, Redlands, CA) and the 2014 North Carolina Department of Transportation road network. Using a road network dataset separate from the online network data providers being tested ensures that all online network data providers must independently allocate the selected point locations to their network, thus favoring no specific snapping or reverse geocoding application (as seen in Figure 1, the impact of small geocoding differences among providers could have significant impact on the calculated route).

2.3.1. Generating candidate locations

A set P of origin and destination points were simulated only along tertiary roads and as such, only locations where it would be possible to begin and end an actual traveled route could serve as candidate locations.[7] The start and end nodes that define each tertiary road segment provide the full set of candidate origin and destination points (set P). The following notation is now introduced:

$p = |P|$:   number of candidate (origin and destination) locations (candidate vertices), P is the set.

$p_c = |P_c|$:   number of candidate (origin and destination) locations (candidate vertices) per county c, P_c is the set.

$q = |Q|$:   number of possible pairs of distances, Q is the set with $|Q| = (\frac{|P|!}{2!(|P|-2)!})$

$n = |N|$:   target number of pairs of distance, N is the set (e.g. 100,000 pairs).

This procedure was implemented in the MATLAB environment and Figure 3(a) provides a flowchart of the procedure used to determine origins, destinations, and resulting OD matrix, while Figures 3(b–d) illustrate the spatial distribution of the selected locations, and the OD matrix for one particular county, respectively. The set of candidate location P is determined using the begin nodes of each road segments in the study area, forming a set $|P|$ of 306,788 vertices (Figure 3(b)), resulting in $|Q| = 47,059,285,078$ potential pairs of distances. Given that MATLAB cannot handle more than 450 million OD entries, we reduced the number of candidate vertices to an acceptable level, reducing the potential vertices to 30,000 ($|P_c| = 300$ per county; there are 100 counties in North Carolina, see Figure 3(c)).

2.3.2. Selecting pairs of distances

We set the number of OD pairs to evaluate to manageable value ($n = 100,000$) given the volume limitations in the proprietary data sources. For instance, without a paid license, Google Maps limits the number of travel estimation requests to 2,500 per day and ArcGIS Online requires a paid subscription in order to use their web services. MapQuest Open does not have a defined limit. Using the full set of potential candidate pairs ($|Q|$, see Figure 3(d)), we sampled $|N| = 100,000$ pairs of distances. To guarantee our samples covered a broad range of potential distances, we defined categories (classes) of distances, K, to use as a stratification scheme in sample selection. For instance, if there would be $k = 10$ categories, each category would have 10,000 samples. This stratified sampling approach is similar to the one conducted to optimize the estimation of the variogram in geostatistics, where samples are optimally located so as to guarantee a certain number of observations within each distance lag (Warrick and Myers 1987). We introduce the following notation for use within the point selection algorithm (see pseudocode in Figure 4).

$k = $ index used to partition the distances into k distinct classes, $k \in K$

$k_{a,b} = $ minimum and maximum distance bounds for each class (e.g. 0-1, 1-2...)

$\frac{|N|}{k} = $ target sample size for each category (for instance when $k = 10$, $\frac{|N|}{k} = 10,000$)

The stratified selection of OD pairs also ensured that the travel distances evaluated span from very short to the range permitted by the study area bounds. The distribution of network distances is different than Euclidean distances (network distances are always greater than Euclidean distances), and instead of $K = 10$ categories we ended up with approximately 14,300 pairs in $K = 7$ distinct distance intervals (total = 100,000 OD pairs):

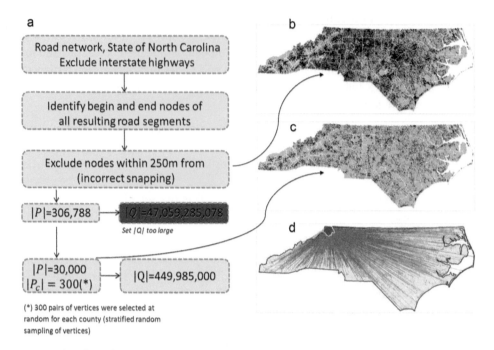

Figure 3. Flowchart depicting the candidate OD selection process in (a). Initial candidate locations are shown in (b) and the reduced set in (c). OD matrix for a single county in figure (d).

0–50 kilometers (km), 50–100 km, 100–150 km, 150–200 km, 200–250 km, 250–500 km and 500–1000 km.[8] The methodology also specified that the number of candidate origin and destination points is equivalent in each of North Carolina's 100 counties. This ensures that the sample travel distances were spatially distributed throughout the study area. Although OD pairs are stratified by distance, the heterogeneity of the street network throughout the study area produces a clustered distribution of OD points gravitating towards higher density, shorter segmented urban road networks.

2.4. VGI user contribution

In MapQuest Open, an XML-formatted response is returned upon completion of the routing request that includes information concerning all road segments (known as a 'way'). For each way, the XML response contains the length of the way and a coordinate indicating the start location. MapQuest Open provides the Nominatim Search API that can reverse geocode the stored list of beginning coordinates for each way, allowing to identify the unique OSM identifier. Using the way identifiers, the OSM dataset provides its own API, which can access the attributes of any feature in the dataset. Passing the identifiers to this API provides the complete history of the OSM object, including the number of users that have contributed to that road segment (Figure 5). As an example, Figure 6 summarizes the number of contributors for each road segment across North Carolina, suggesting that higher number of updates on each road segment are typically found in urban areas.

```
#Define the lower and upper distance bound of each class k
Define kₐ and kᵦ ∀k

#set the number of observations per bin
numObsPerBin = zeros(|K|,1)

#set an empty list keeping track of the pairs already selected
visitedPairs = []

#set i as the counter for the current sample size
i=0

while i < |N|:
    pass = 0
    while pass == 0:
        #randomly select a distance in the set Q;
        #if it has been chosen already, generate a new one
        flag = 0
        while flag == 0:
            e = rand*|Q|
            if e in visitedPairs:
                flag = 0
            else:
                flag = 1
                visitedPairs = [visitedPairs; e]
                compute e    #e: euclidean distance

                #finding distance class where "e" will belong
                for k in |K|:
                    if ka<e<kb:
                        if numObsPerBin(k,1)<(|N|\k):
                            numObsPerBin(k,1) +=1
                            pass = 1
                    else:
                        k+=1

    i+=1
```

Figure 4. Pseudo-code outlining the algorithm used to uniformly select OD pairs in stratified distance intervals.

2.4.1. *Average number of contributors*

A distance weighted contributor average C_a is calculated for each route, using the number of contributors for each road segment that comprise this route. The average is proportional to the length of known[9] road segments:

$$C_a = \sum \left(\frac{d_w}{d_t} * c_w \right); \text{ where } 0 > C_a > \infty \tag{3}$$

With d_w the distance of each road segment used in the total route calculation, d_t the total distance of all sampled road segments from the total route calculation and c_w the number of contributors for each road segment used in the total calculation (Figures 7(a–c) gives a visual illustration on how the average number of contributors is computed based on an optimal route (selected path, in red) and different road segments for which the number of contributors is known.

3. Results

Travel estimates and average contributors were estimated for the simulated routes in the summer of 2015. A total of 99,910 out of the 100,000 routes selected were viable for

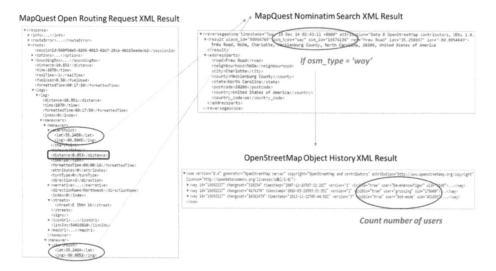

Figure 5. Steps required to obtain the number of users for each OSM road segment.

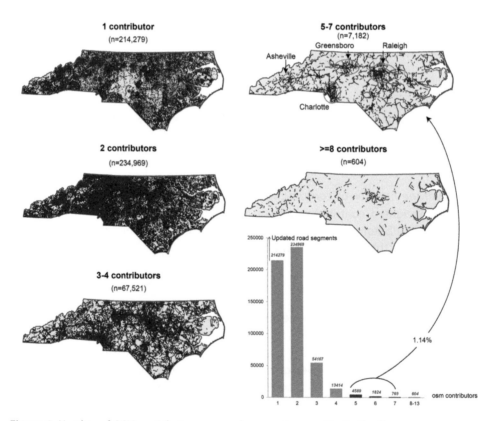

Figure 6. Number of OSM contributors per road segment across North Carolina.

analysis; MapQuest Open could not complete 90 OD routes, mostly due to incomplete VGI data resulting in unconnected sections of the OSM road network.[10]

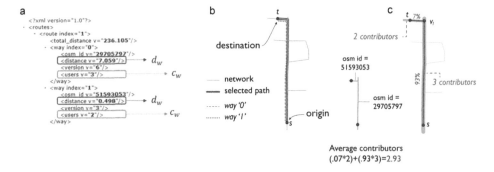

Figure 7. XML output for an optimal route and its different elements (ways) in (a), average number of contributors based on an optimal route (selected path) in (b) and the number of contributors for each road segment the selected path traverses in (c).

3.1. Comparison of distance estimations across providers

High correlation values and relatively low uncertainty among providers are found in Figures 8(a-c). Relative comparisons of these values indicate patterns of (dis)agreement in connection to total travel distance. Travel distance uncertainty can also be visualized (by the tightness of points) in the scatter plots shown in Figures 8(a-c); the less variability in estimated travel times, the less uncertainty among the two providers being compared.

The slope in these graphs provides some indication as to why; in comparison to Google Maps, ArcGIS Online tends to overestimate travel distance. Conversely, MapQuest Open tends to underestimate travel distances. The graphs show similar outliers at shorter travel distances. Some of this variation can be explained by different routing choices. Several of these outliers have an origin or destination on the barrier islands of the North Carolina coast, resulting in a different travel routing decision between providers, ultimately increasing travel impedance.[11] For the scatter plots in Figures 8(d)–9(f), Google Maps was selected as the reference travel distance to provide a consistent scale for comparison. While the absolute uncertainty is represented in the Figures 8(a)–9(c), the percent difference plots in Figures 8(d)–9(f) illustrate relative uncertainty by travel distance. Overall, the percent difference values tend to be larger at shorter distances and decrease at longer travel distances. While this is not unexpected, given that short distances will be much more sensitive to small differences in the calculation, the plots do demonstrate some deviation from a uniform decrease, e.g. near 150 miles in Figure 9(f).

3.2. VGI user contribution as a measure of uncertainty

The distance weighted average of user contributors was estimated for all 99,910 routes. The average number of contributors for each route was 3.27 and only 95 routes had an average of 9 or more contributors. For each route, the average user contribution was plotted against the average agreement between MapQuest Open and Google Maps (Equation 2), expressed in percentage. The percent disagreement is thus the average (Equation 2) of all 3% difference calculations among network providers. Next, user averages were then allocated into discrete intervals for analysis. User averages ranging

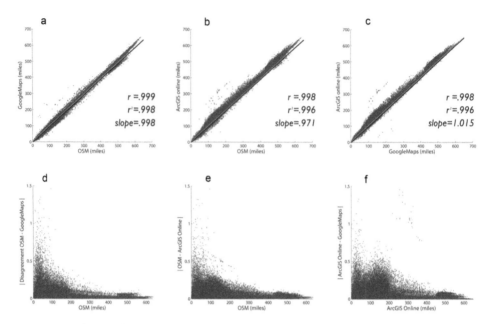

Figure 8. Correlation plots with 1:1 line in red (a–c) and percent differences (d–f) for network distance estimates.

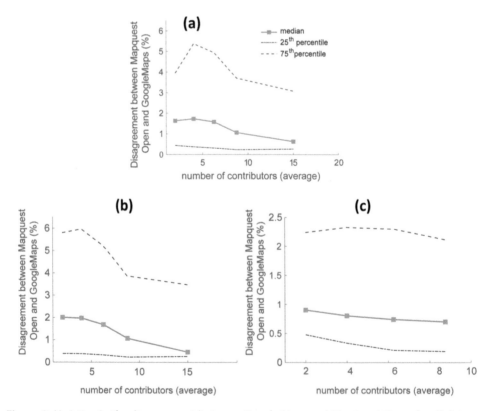

Figure 9. Variation in the disagreement between Google Maps and MapQuest Open, for all distances in (a), less than 250 miles in (b) and greater than 250 miles in (c).

less than 1.01 were categorized as having one contributor. The lower and upper bounds of the remaining categories were determined so that they would include a sufficient number of observations ($n > 25$), ending with the following categories: {1}, {1.01 to 3}, {3.01 to 5}, {5.01 to 7.5}, {7.51 to 10} and {10.01 to 20}. For each category, we reported the median value, the 25th and 75th percentile (Figure 9(a)). We then decomposed the results for short (distances less than 250 miles) and longer routes (distances greater than 250 miles) and present them in Figures 9(b,c) respectively.

From Figure 9(a), we observe an initial increase in percent disagreement when the number of contributors is small; this peak coincides with the largest sample size intervals as well as the mean number of users across the entire sample (3.27). Importantly, the level of disagreement decreases as the number of contributors' increases. The variation (difference between the 75th and 25th percentile) is lower among the higher average user intervals. Comparing Figures 9(b,c), we note a lower level of disagreement between MapQuest Open and Google Maps at longer distances (the range of the Y-axis is much lower in Figures 9(c)), but that the number of contributors has a greater likelihood of reducing disagreement for smaller travel than longer ones. Finally, very few long routes had a large number of contributors.

4. Discussion and conclusions

Online network data providers emphasize the power of web services as a tool for spatial analysis, yet the limitations set by proprietary providers pressure the need for an open source, online alternative. The primary example of such a dataset, OSM, has recently been the focus of much data quality studies to identify its reliability for research. While most data quality research has evaluated VGI's positional (geometry) and attribute accuracy in reference to a proprietary or authoritative dataset, the practicality of this approach may not be suitable for a dynamic geographic dataset such as OSM. Examining travel impedance estimates as described in this paper offers a novel assessment of VGI data quality, given that these estimates can be influenced by multiple categories of data accuracy (i.e. positional accuracy, attribute accuracy and completeness). While the method provide here requires the use of commercial online network data, this scalable and automated approach can be readily used with other reference data sources to indicate areas of travel impedance uncertainty. In turn, these uncertainty measurements can lead to a more spatially focused review of OSM data quality where it is warranted.

Results of this study indicate that travel impedance estimates calculated using Google Maps, ArcGIS Online, and MapQuest Open (which is based on OSM data) have strong agreement. Travel impedance correlation coefficients (r) calculated between this set of providers was greater than 0.98. We found that high disagreement exceptions were likely due to underlying topological characteristics that make any network travel estimation difficult and uncertain. For example, high disagreement in travel could be the result of differing routes altogether, rather than errors in either dataset. Possible reasons why ArcGIS Online tended to overestimate Google Maps could include slight differences in the coding of the heuristic used for the shortest path algorithm, rounding of travel estimates, and the fact the Google Maps has a more up-to-date and complete road network as opposed to ArcGIS Online. Yet, these results are salient to our overall findings, as they may be the result of incomplete data or topological errors that contribute to the uncertainty in estimates.

The low uncertainty within travel impedance estimates among several online network data providers indicates that MapQuest Open is a viable option for routing and navigation purposes within the study area. In particular, MapQuest Open presents greater agreement with other network data providers at longer distances. This does not imply that MapQuest Open is a reliable service for all applications. While the dataset is comprehensive enough to provide a routing calculation for almost every OD pair, any uncertainty along the selected routes may be unacceptable if applied for anything greater than personal use, e.g., for research purposes.

Previous assessments of OSM data have largely been conducted across Europe. The development of the US OSM dataset may be fundamentally different from other parts of the world; specifically, the OSM road network in the USA is extensively derived from the US TIGER Census files, a reference dataset for many geospatial data quality assessments. Because of this, much of the road network may not have needed revisions from OSM contributors. Yet, even with this caveat, we found evidence that suggests *Linus' Law* held in this situation as the percentage disagreement among OSM and the other network distance providers decreased as the number of contributors grew. This was especially apparent when travel distances were relatively long, which requires identifying the travel distance along numerous, disparate road segments to accurately measure. While we cannot evaluate the deviation from the 'true' shortest distance, the results suggest that an increased number of contributors does bring the OSM results nearer to data sources often considered as more accurate. As such, this paper expanded the literature concerning the applicability of *Linus's Law* for VGI datasets.

We see several avenues for future research. First, because this research was the first application of an OD matrix for VGI data quality assessment, we did not intend to explain the cause for travel impedance estimate agreement among online network data providers. Instead, it was the intention of this paper to discover trends in the patterns of disagreement and investigate the viability of this assessment method using a VGI-based online network dataset. Second, this research focused on travel distance as the indicator for travel impedance estimate uncertainty. Future research should test whether the findings of this research are also valid for time as travel impendence. Indeed, each provider may have different algorithms or rules to account for traffic and en-route condition (such as congestion). Accurate travel time estimates are particularly important for emergency medical services and fire rescue operations. Travel time is affected by traffic, which varies with time of day and day of the week. Although Google Maps allowed a travel requests to include specifics on time of day and day of the week, MapQuest Open and ArcGIS Online did not allow to define such parameters at the time the analysis was conducted. Third, more efforts should investigate whether the findings found in this paper can be replicated across different landscapes with varying degrees of urbanicity and in other countries that have witnessed a rapid increase in OSM coverage. Along those lines, the applicability of *Linus's Law* to the U.S. OSM dataset should be expanded, increasing the size of the analysis to verify if our findings can be replicated elsewhere. Finally, it would be beneficial to verify the amount of overlap in the routes selected among different providers, such as in Ludwig *et al.* (2011). Studying the exact route selection could explain why routes may differ in travel impedance estimates.

Notes

1. Google Maps uses various datasets, including municipality and county maps, satellite imagery and through their imaging vehicles. Google's dataset has a lot of public input from their acquisition of Waze. Google has built proprietary technology to fuse these disparate data sources to generate the most accurate maps. Google also employs individuals who manually correct the information collected from these different sources.
2. https://wiki.openstreetmap.org/wiki/TIGER.
3. In rural areas, authoritative data imports may be considered 'good enough' whereas if there was no data at all, users might feel more obligated to provide road network information instead of contributing extemporaneous attributes that do not improve core data quality.
4. In the case of linear elements, many of the road segments with greater than five contributors are generally highways or streets located in populated regions.
5. ArcGIS online source for their road network data is provided by HERE Technologies, which used to be called Navteq (competitor to TeleAtlas and of Google Maps).
6. Because these routing web services are proprietary, the exact routing algorithm each service uses to complete a shortest path is not known. However, due to the popularity and robustness of Dijkstra's algorithm, it is reasonable to assume all the services use this algorithm or a derivative. Under this assumption, differences in travel impedance calculations between providers should not vary significantly in this respect.
7. When a coordinate is entered in an on-line routing application, the provider automatically snaps it to the closest available road vertex or node. Interstates and limited-access highways were removed from the network dataset that was used to generate candidate locations. In addition, as a final assurance that no locations would coincide with the begin- or endnode of an interstate or a highway ramp, any nodes within 250 m of the previously removed roads were removed from the candidate point set.
8. Because few travel distances approach the maximum travel distance across the state, it is necessary to increase the range of the category intervals for the longer distances to accomplish an equally stratified sample.
9. We use reverse geocoding to identify road segments that make up a route. However, this procedure does not systematically return a response. Therefore, it may happen that some road segments part of the route are not accounted for. As such, we only calculate the user average on the segments that we could account for.
10. The distribution of calculated network distances departs from the uniform categories used to select the OD pairs, since network distances are consistently greater than the straight-line distance estimates.
11. At the time of the experiment, Google Maps incorporated ferry information between coastal cities of North Carolina, while ArcGIS Online did not.

Disclosure statement

No potential conflict of interest was reported by the authors.

ORCID

Eric M. Delmelle ⓘ http://orcid.org/0000-0002-5117-2238
Paul L. Delamater ⓘ http://orcid.org/0000-0003-3627-9739

References

Apparicio, P., *et al.*, 2008. Comparing alternative approaches to measuring the geographical accessibility of urban health services: distance types and aggregation-error issues. *International Journal of Health Geographics*, 7 (1), 7. doi:10.1186/1476-072X-7-7

Arsanjani, J.J., *et al.*, 2015. OpenStreetMap in GIScience. *In: Lecture notes in geoinformation and cartography*. Springer International Publishing.

Bakillah, M., *et al.*, 2014. Fine-resolution population mapping using OpenStreetMap points-of-interest. *International Journal of Geographical Information Science*, 28 (9), 1940–1963. doi:10.1080/13658816.2014.909045

Berke, E.M. and Shi, X., 2009. Computing travel time when the exact address is unknown: a comparison of point and polygon ZIP code approximation methods. *International Journal of Health Geographics*, 8 (1), 23. doi:10.1186/1476-072X-8-23

Boscoe, F.P., Henry, K.A., and Zdeb, M.S., 2012. A nationwide comparison of driving distance versus straight-line distance to hospitals. *The Professional Geographer*, 64 (2), 188–196. doi:10.1080/00330124.2011.583586

Casas, I., Delmelle, E., and Delmelle, E.C., 2017. Potential versus revealed access to care during a dengue fever outbreak. *Journal of Transport & Health*, 4, 18–29. doi:10.1016/j.jth.2016.08.001

Ciepluch, B., *et al.*, 2010. Comparison of the accuracy of OpenStreetMap for Ireland with Google Maps and Bing Maps. *In: Paper read at proceedings of the ninth international symposium on spatial accuracy assessment in natural resources and environmental sciences*, July 20–23, Leicester, England.

Corcoran, P., Mooney, P., and Bertolotto, M., 2013. Analysing the growth of OpenStreetMap networks. *Spatial Statistics*, 3, 21–32. doi:10.1016/j.spasta.2013.01.002

Delamater, P.L., *et al.*, 2012. Measuring geographic access to health care: raster and network-based methods. *International Journal of Health Geographics*, 11 (1), 15. doi:10.1186/1476-072X-11-15

Delmelle, E.M., *et al.*, 2013. Modeling travel impedance to medical care for children with birth defects using geographic information systems. *Birth Defects Research Part A: Clinical and Molecular Teratology*, 97 (10), 673–684. doi:10.1002/bdra.v97.10

Dijkstra, E.W., 1959. A note on two problems in connexion with graphs. *Numerische mathematik*, 1 (1), 269–271. doi:10.1007/BF01386390

Dony, C.C., Delmelle, E.M., and Delmelle, E.C., 2015. Re-conceptualizing accessibility to parks in multi-modal cities: a Variable-width Floating Catchment Area (VFCA) method. *Landscape and Urban Planning*, 143, 90–99. doi:10.1016/j.landurbplan.2015.06.011

Girres, J.F. and Touya, G., 2010. Quality assessment of the French OpenStreetMap dataset. *Transactions in GIS*, 14 (4), 435–459. doi:10.1111/j.1467-9671.2010.01203.x

Goetz, M. and Zipf, A., 2012. Using crowdsourced geodata for agent-based indoor evacuation simulations. *ISPRS International Journal of Geo-Information*, 1 (2), 186–208. doi:10.3390/ijgi1020186

Goodchild, M.F., 2007. Citizens as sensors: the world of volunteered geography. *GeoJournal*, 69 (4), 211–221. doi:10.1007/s10708-007-9111-y

Goodchild, M.F. and Li, L., 2012. Assuring the quality of volunteered geographic information. *Spatial Statistics*, 1, 110–120. doi:10.1016/j.spasta.2012.03.002

Griffith, D.A., 2018. Uncertainty and context in geography and GIScience: reflections on spatial autocorrelation, spatial sampling, and health data. *Annals of the American Association of Geographers*, 1–7.

Gutiérrez, J. and García-Palomares, J.C., 2008. Distance-measure impacts on the calculation of transport service areas using GIS. *Environment and Planning B: Planning and Design*, 35 (3), 480–503. doi:10.1068/b33043

Haklay, M., 2010. How good is volunteered geographical information? A comparative study of OpenStreetMap and ordnance survey datasets. *Environment and Planning. B, Planning & Design*, 37 (4), 682. doi:10.1068/b35097

Haklay, M., *et al.*, 2010. How many volunteers does it take to map an area well? The validity of Linus' law to volunteered geographic information. *The Cartographic Journal*, 47 (4), 315–322. doi:10.1179/000870410X12911304958827

Hochmair, H.H., Zielstra, D., and Neis, P., 2015. Assessing the completeness of bicycle trail and lane features in OpenStreetMap for the United States. *Transactions in GIS*, 19 (1), 63–81. doi:10.1111/tgis.2015.19.issue-1

Jilani, M., *et al.*, 2013b. Automated quality improvement of road network in OpenStreetMap. *In: Paper read at Agile Workshop (action and interaction in volunteered geographic information)*, Leuven, Belgium.

Jilani, M., Corcoran, P., and Bertolotto, M., 2013a. Multi-granular street network representation towards quality assessment of OpenStreetMap data. *In: Paper read at proceedings of the Sixth ACM SIGSPATIAL international workshop on computational transportation science*, November 05 - 08, ACM, Orlando, FL.

Jones, S.G., *et al.*, 2010. Spatial implications associated with using Euclidean distance measurements and geographic centroid imputation in health care research. *Health Services Research*, 45 (1), 316–327. doi:10.1111/j.1475-6773.2009.01044.x

Keßler, C. and de Groot, R.T.A., 2013. Trust as a proxy measure for the quality of volunteered geographic information in the case of OpenStreetMap. *In: Geographic information science at the heart of Europe*. Switzerland: Springer International Publishing, 21–37.

Kirby, R.S., Delmelle, E., and Eberth, J.M., 2017. Advances in spatial epidemiology and geographic information systems. *Annals of Epidemiology*, 27 (1), 1–9. doi:10.1016/j.annepidem.2016.12.001

Klonner, C., *et al.*, 2015. Updating digital elevation models via change detection and fusion of human and remote sensor data in urban environments. *International Journal of Digital Earth*, 8 (2), 153–171. doi:10.1080/17538947.2014.881427

Ludwig, I., Voss, A., and Krause-Traudes, M., 2011. A comparison of the street networks of Navteq and OSM in Germany. *In: Advancing geoinformation science for a changing world*. Berlin: Springer, 65–84.

Mooney, P., Corcoran, P., and Winstanley, A.C., 2010. Towards quality metrics for OpenStreetMap. *In: Paper read at Proceedings of the 18th SIGSPATIAL international conference on advances in geographic information systems*, November 03–05, ACM, San Jose, CA.

Murray, A.T., 2010. Advances in location modeling: GIS linkages and contributions. *Journal of Geographical Systems*, 12 (3), 335–354. doi:10.1007/s10109-009-0105-9

Murray, A.T. and Tong, D., 2009. GIS and spatial analysis in the media. *Applied Geography*, 29 (2), 250–259. doi:10.1016/j.apgeog.2008.09.002

Neis, P., 2015. Measuring the reliability of wheelchair user route planning based on volunteered geographic information. *Transactions in GIS* 19.2: 188–201.

Neis, P. and Zielstra, D., 2014. Recent developments and future trends in volunteered geographic information research: the case of OpenStreetMap. *Future Internet*, 6 (1), 76–106. doi:10.3390/fi6010076

O'Sullivan, S. and Morrall, J., 1996. Walking distances to and from light-rail transit stations. *Transportation Research Record: Journal of the Transportation Research Board*, 1538 (1538), 19–26. doi:10.1177/0361198196153800103

Peleg, K. and Pliskin, J.S., 2004. A geographic information system simulation model of EMS: reducing ambulance response time. *The American Journal of Emergency Medicine*, 22 (3), 164–170.

Phibbs, C.S. and Luft, H.S., 1995. Correlation of travel time on roads versus straight line distance. *Medical Care Research and Review*, 52 (4), 532–542. doi:10.1177/107755879505200406

Racine, E.F., *et al.*, 2018. Accessibility landscapes of supplemental nutrition assistance program– Authorized stores. *Journal of the Academy of Nutrition and Dietetics*, 118 (5), 836–848. doi:10.1016/j.jand.2017.11.004

Schmitz, S., Zipf, A., and Neis, P., 2008. New applications based on collaborative geodata—the case of routing. *In: Paper read at Proceedings of XXVIII INCA international congress on collaborative mapping and space technology*, Gandhinagar, Gujarat.

Shahid, R., *et al.*, 2009. Comparison of distance measures in spatial analytical modeling for health service planning. *BMC Health Services Research*, 9 (1), 200. doi:10.1186/1472-6963-9-200

Socharoentum, M. and Karimi, H.A., 2014. A comparative analysis of routes generated by web mapping APIs. *Cartography and Geographic Information Science* 42.1: 33–43.

Soden, R. and Palen, L., 2014, May 27–30. From crowdsourced mapping to community mapping: the post-earthquake work of OpenStreetMap Haiti. *In: Paper read at COOP 2014-proceedings of the 11th international conference on the design of cooperative systems*. Nice (France): Springer International Publishing.

Wang, F. and Xu, Y., 2011. Estimating O–D travel time matrix by Google Maps API: implementation, advantages, and implications. *Annals of GIS*, 17 (4), 199–209. doi:10.1080/19475683.2011.625977

Warrick, A.W. and Myers, D.E., 1987. Optimization of sampling locations for variogram calculations. *Water Resources Research*, 23 (3), 496–500. doi:10.1029/WR023i003p00496

Widener, M.J., Metcalf, S.S., and Bar-Yam, Y., 2011. Dynamic urban food environments: a temporal analysis of access to healthy foods. *American Journal of Preventive Medicine*, 41 (4), 439–441. doi:10.1016/j.amepre.2011.06.034

Zhang, J. and Goodchild, M.F., 2002. *Uncertainty in geographical information*. New York, NY: CRC press.

Zook, M., *et al.*, 2010. Volunteered geographic information and crowdsourcing disaster relief: a case study of the Haitian earthquake. *World Medical & Health Policy*, 2 (2), 7–33. doi:10.2202/1948-4682.1069

A network approach to the production of geographic context using exponential random graph models

Steven M. Radil ⓘ

ABSTRACT
The notion of context continues to be both an enduring rationale and empirical problem for addressing human agency for geographers. Despite its centrality to geographic scholarship, context has largely been an abstraction in geography with relatively little effort to either clarify what it means or how to formally operationalize it for research purposes. When context has been formally addressed, it has primarily emphasized either impacts on agency at a specific scale through a reliance on a place-based interpretation. This paper takes up the issue of context by developing a multi-scalar theoretical framework that is suited for use with social network-based statistical models called exponential random graph models or ERGMs. The theory of context emphasizes the importance for both geographic and social contexts for agency while also recognizing place specific and larger scale influences. Using network data about World War I, a series of ERGMs are developed to demonstrate the importance of multiple types of contexts to the observed outcomes. The approach used in this paper reinforces the old truism that context matters by demonstrating it empirically. Most importantly, this paper illustrates the value of continued engagement with a wider spectrum of the theories of how and why context matters.

Introduction

The notion of context continues to be both an enduring rationale and empirical problem for addressing human agency for geographers. Indeed, many of the prominent debates and shifts in social theory in geography, such as the relational turn, can be understood as efforts to better contextualize the unfolding of human agency. This is undoubtedly because many of geography's most important concepts, such as place, scale, and location, are themselves efforts to more properly situate human action, intention, and outcomes. Despite its centrality to geographic scholarship, context has largely been an abstraction with relatively little effort to either clarify what it means or how to formally operationalize it for research purposes. When context has been formally addressed, it has primarily emphasized either impacts on agency at small or localized spatial scales through the peculiarities of place (e.g. Agnew 1987), relied on limited geometric or distance-decay informed notions of 'neighbors' to formalize spatial dependence for spatial analysis (e.g. Leenders and Roger 2002), or been concerned with problems of assessing the micro-scale contexts of mobile individuals (e.g. Kwan 2012).

Although the importance of context to the discipline of human geography is unquestioned, the discussion of what context is and why it matters remains underdeveloped in geographic thought (Grossberg 2013). This is a missed opportunity given the persistent retheorizations of other core concepts in geography (space, place, scale). Context, then, is in need of an intervention, especially for use in quantitative research where context is either conspicuously absent or incorporated in a relatively simple or incomplete fashion (Flint et al. 2009). This is a particularly pressing need given that quantitative research in geography is often caricatured by its critics as interested in developing and testing geographically invariant universal laws and therefore opposed to critical geographic inquiry (see Barnes 2009). In other words, quantitative work is often seen as intentionally acontextual.

As noted by Barnes and others, such critiques of quantitative research are anchored in the epistemological debates of the late 1960s and early 1970s rather than a clear-eyed view of contemporary quantitative geography (see also Sheppard 2011). Fotheringham et al. (2004, p. 11) observed that quantitative geography now emphasizes differences across space rather than similarities and concentrates on identifying 'local exceptions rather than the search for global regularities.' Such exceptions to global regularities might be interpreted as the outcomes of contextualized processes but this is not quite the same as an effort to explicitly make context central to quantitative research in geography. More is needed than just oblique references to context, especially since operationalizing context for quantitative work may have the potential to continue the diffusion of geographic ideas in social science more broadly (Warf and Arias 2008).

This paper takes up the dual challenge of theorizing context and operationalizing it for quantitative research in geography. This is done with the use of social network analysis in general and a family of network-based models called Exponential Random Graph Models (or ERGMs) in particular. The paper first reviews issues connected to the concept of context in geography to develop a tripartite theory of context that incorporates (1) actor- or place-characteristics, (2) relationships between actors or places, including spatial relationships, and (3) larger systemic properties that are not reduceable to any of the above. ERGMs are then introduced along with a dataset that is used to demonstrate the issues involved. A series of models are developed that correspond to different elements of context which shows that the best performing model is the one that incorporates all three proposed elements of context. The paper concludes with a discussion of the benefits to matching theories of context with appropriate modeling techniques for quantitative inquiry.

Context in geography

Context matters. If there's a single phrase that could sum up the differences between contemporary geography and other social science fields, this is it. Yet context has been an undertheorized concept in geography (Flint and Dezzani 2017; Goodchild 2018) which poses a particular problem for geographical analyses that presumably seek to take context seriously. While other core geographical concepts have been subject to a great deal of theoretical scrutiny (scale, place, space), more effort is needed to clarify what it is geographers mean by context and how their scholarship can demonstrate its impact analytically. Context may then be geography's primary (but likely not only) 'fuzzy concept,' a term coined by regional studies scholar Ann Markusen (1999, p. 871) to refer to a concept that 'possesses two or more alternative meanings and thus cannot be reliably identified or applied by different readers or scholars.'

Markusen (1999) argued that the use of fuzzy concepts typically involves an abstraction of a process that tends to gloss over or obscure altogether actions by specific actors and institutions. In this way, she claims that their use tends to situate either important actors (and their agency) or key structures outside of the scope of an analysis, typically by invoking 'contingencies' to explain an effect that is actually produced by something/one more specific. Perhaps most importantly, Markusen claimed the reliance on fuzzy concepts has had the effect of diminishing the impact of geographic scholarship as it has led to a devaluation of evidence and method, reduced replicability, and limited the potential for scholars to inform action within a society. This is problematic for geographers and those outside the discipline that may look to geography to demonstrate how and why context matters.

The issues associated with fuzzy concepts are not exclusive to either qualitative or quantitative work. However quantitative work has a clear tradition of trying to identify uncertainty and ambiguity; such methods, therefore, tend to be more transparent which can, paradoxically, prompt methodological critique and debate in a way that is largely absent from qualitative work (Fotheringham *et al.* 2004; see also Radil and Flint 2013, 2015, Verweijen and van Meeteren 2014 for an example of such a debate). Further, scholars working from the intertwined GIScience and spatial analytic traditions in geography (Goodchild and Haining 2004) are well positioned to take up the challenges of clarifying context. Goodchild (2018, p. 1) argues for more interest in context in GIScience by pointing out that geographers typically try to understand a phenomenon through reference to its location (site) and surroundings (situation or environment) and that GIScience can be used to understand both: 'much of the richness of geographic science stems from the ready availability of information on context.' Context then, is both a key challenge and rationale for GIScience even though, as Grossberg (2013, p. 32) put it, context 'is almost never defined, and even more rarely theorized.'

Fuzzy concepts are often useful because they are ideas in a state of development. This implies that a fuzzy concept is an immature one and maturity will necessarily involve scrutiny, debate, and eventually the imposition of more precision into the concept. Context is an important concept to consider in this line of critique as it is often the idea that separates geography from other social science scholarship (see Ethington and McDaniel 2007 for an example drawn from political geography and political science). To the extent that geographers have taken on the task of clarifying context, it's been done largely through the use of another concept, that of place (e.g. Massey 1993). More specifically, context and contextual effects on human behavior are typically approached in geography by arguing for the salience of a place-based understanding of agency (e.g. O'Loughlin 2000).

Agnew's (1987) formalization of place has been widely used to consider context perhaps because it was developed with regard for explaining election outcomes. Agnew (1987) asserts that variable combinations of the economic, institutional, and experiential elements of a place serve to mediate political behavior within places themselves. Election outcomes in a place then are partly a function of the combination of these elements on individual decision-making and help to explain differing outcomes among voters that might otherwise be thought to share interests due to similar politico-social attributes (such as race or gender) or shared group membership (such as middle-class voters). In Agnew's words, places are where 'lives are lived, economic and symbolic interests are defined, information from local and extra-local sources is interpreted and takes on meaning' Agnew (1987, p. 2).

This interpretation of why place matters have been helpful in clarifying how agency is formed and occurs within places, even if the exercise of agency involves aspects which lie beyond the place itself or encompass multiple places simultaneously (see also Pred 1984 for a similar interpretation on place as the primary setting of human agency). Helpfully, place-based studies are typically focused on specific actors (whether individuals, social collectives, or formal institutions) whose activities can be directly observed within a relatively small setting. And yet care is needed by equating context exclusively with the 'local' scale or by equating place with overly small- or micro-scale settings. By privileging the local, a place-based approach to context can risk overlooking or excluding agency at other, often larger, spatial scales (see Massey 1997, Grossberg 2013). This is a point made clearly by Agnew (1996, p. 132) as part of the well-known 'context' debate with political scientist Gary King: '[C]ontext refers to the hierarchical (and non-hierarchical) "funneling" of stimuli across geographical scales or levels to produce effects on politics and political behavior. These effects can be thought of as coming together in places where micro (localized) and macro (wide-ranging) processes of social structuration are jointly mediated.' Put another way, any approach to context that treats it as associated with or manifest through a single spatial scale will miss a significant part of the story.

For King's (1996) part in this debate, addressing context was simply an exercise in being able to observe the correct variables for regression models, either of the area/spatial unit in which political actors were located, or of the actors themselves. What geographers call 'contextual effects' were just then the byproduct of unobserved (and perhaps unobservable) variables. Therefore, in King's (1996, p. 160), the purpose of geography should be to 'try as hard as possible to make context not count' by identifying and measuring the variables that may create contextual effects. This approach to context (if it may be said to be an approach at all) argues that there may be very localized contextual impacts on voter behavior but that these may be captured simply by observing the appropriate variable(s).

King's stand on context has been critiqued by geographers (Agnew 1996) and also by some political scientists. For example, Ethington and McDaniel (2007, p. 134–135) take issue with King's approach, calling it 'weak contextualism' and arguing that it is essentially interested in only very small scale effects produced in microscale settings: 'context is conceived entirely as based around individuals and the influence of their friends and family.' Ethington and McDaniel also assert that a more productive direction is found in the work of scholars interested in relational networks of social interaction than span both the social scale of individuals and the localized spatial scales of the home/workplace/etc. This is an obvious bridge to the growing body of work by geographers interested in social networks and how to combine them with other forms of geographic data (e.g. Radil et al. 2010; Sui and Goodchild 2011).

Theorizing context

The Agnew-King exchange was an important moment where context was subjected to the type of process Markusen was concerned about: open discussion of an abstraction that moved it toward more clarity and analytic usefulness. Given that this exchange was nearly 20 years ago, it may seem as though there has been little theorizing in geography since. However, a close look at some of the literature in geography and GIScience turns up some notable exceptions. For instance, Flint's (2002) study of the electoral geography of the Nazi party in pre-World War II Germany argued that the use of spatial statistical models that accommodate the property of spatial dependence captures some elements of context, particularly the notion that social relations may extend beyond specific locations and therefore serve to define regions of contextual similarity (Flint 2002; see also O'Loughlin *et al.* 1994). Such regions may be expected to produce spatially hetero-geneous political outcomes that may substantively differ from any global patterns present. This approach aimed to capture context at a particular scale, which the author called the 'regional scale,' by creating a new variable (the spatial lagged mean of voting outcomes) to include as an explanatory variable in a spatial regression model.

This approach to context, common in the spatial analysis literature, is limited for a variety of reasons. First, it requires the analyst to specify the form and extent of connections between locations to create the spatial lagged mean. In practice, this is often done in an offhand or atheoretical way, such as by connecting only the nearest units through some measure of spatial proximity or contiguity or by specifying a globally applicable distance decay effect to connect units within a given study area (Leenders and Roger 2002, Tita and Radil 2011). Of course, this is a simple task in most GIS software but 'context' may then be partly a function of the choices made by the analyst and may not reflect the actual relationships between actors that create and reflect context. Further, model results (meaning the point estimates of the coefficients of other variables) can be sensitive to these specifications and an offhand approach can shape the model's findings and therefore the analyst's conclusions. Second, as the incorporation of context happens only at a particular scale (defined by the form of the connections specified by the analyst), it is therefore unlikely to reflect the multiscalar connections so important to Agnew's formulations of context Agnew (1987, 1996).

Flint and others have taken up this issue with the aim of developing alternate approaches to context that lend themselves to systematic quantitative analysis. These efforts have incorporated spatial analysis with social network theory and techniques to field spatial distance or contiguity with other types of social relationships to broaden Goodchild's (2018) formulation of situation. For instance, Flint *et al.* (2009) argued in favor of seeing context through the lens of 'embeddedness' to better consider the decision-making context of individual states during periods of war in the interstate system. Embeddedness, initially introduced to geography through the work of sociolo-gist Granovetter (1985, p. 487), argued that human agency is always embedded in 'ongoing systems of social relations.' Flint et al.'s argument was that physical proximity in geographic space also needed to be considered alongside social, economic, or political relations. This approach was extended to other issues, including urban violence (Radil *et al.* 2010) and civil war (Radil and Flint 2013).

Flint and his colleagues drew heavily on Hess's (2004) effort to retheorize embeddedness for their own work. According to Hess, a key deficiency of embeddedness as typically put into practice by geographers was that it was too localized, or in his phrasing, overterritorialized. This had the effect of overlooking some relationships that helped to constitute context but that may be conceptualized at other geographic scales. Instead Hess offered a tripartite approach to embeddedness that was meant to capture multiple dimensions (and therefore, scales) of context. The first, which he called social embeddedness, focused on the character of actors which is partly analogous to a compositional approach to context that emphasizes specific attributes of actors and of the places they inhabit. The second, network embeddedness, emphasized the actual and multiple relationships that create connections between actors. The third and last, territorial embeddedness, reflected a concern for the spatial arrangement of actors and the effects of distance on interaction (see also Flint and Dezzani 2017).

Grossberg (2013) also argued in favor of a tripartite approach to context, which he called 'three modes of contextualization' that drew on Deleuze and Guattari's efforts to theorize space as relational and active. The first mode, milieu or location, reflected Deleuze and Guattari (1987, p. 313) concern for the 'composing elements and composed substances.' To Grossberg (2013, p. 37), these elements make up an 'assemblage of political, economic, social and cultural practices, structures and events' that comprise a milieu; in a way, this too is a compositional approach to context. The second, territory or place, is concerned with the complex relationships between people within a place and between sets of places. As with Hess, webs of social relations comprise a key approach to context. Third and last, Grossberg proposed a notion of region or epoch to address 'the forms of existence – ways of being in space-time – that are possible and that constitute the contingent conditions of possibility of milieus and territories, locations and places' (Grossberg 2013, p. 38). This is concerned with emergent aspects of the broader systems formed by agency. These aspects can change over time and space but serve to impact actors and places without being reduceable to them.

When taken alone, many of the approaches to context only capture partial aspects of Hess's or Grossberg's more holistic approaches (see Table 1 below). For example, Goodchild's (2018) notion of context as composed of site and situation (location and neighborhood) combines some of King's weak context (compositional variable of a place) with Flint's notion of scalar connections (a fixed scale of interactions beyond the place) but does not clearly reflect a concern for larger systems. However, when reviewed together, these various conceptualizations yield useful lessons for a holistic theory of context. First, they emphasize that context is inherently multi-scalar and that place specific context is not just a function of actors located within the place itself. Second, they recognize that the spatial configurations of actors are important but particularly so when fielded alongside the actual relationships that tie actors in space. Third, they draw attention to relational approaches to context which underpin the methodologies associated with network analysis. In this sense, metaphors about networks as 'webs of relations' and so on have the potential to yield to specific methodologies that have much in common with some types of spatial analysis (see Radil et al. 2010 for a discussion of the importance of networks in geographic thought).

What is then needed are methodologies that allow context to better align with the understandings offered by these theoretical efforts. This is a contribution that GIScience

Table 1. Comparison of theorizations of context. All involve some limitations and none capture all the elements of the conceptualizations by Hess (2004), Grossberg (2013), or that are proposed in this paper.

Concept	Scale	Data type	Strengths	Limitations
Actor attributes (compositional context, King 1996)	Individual/ actor focus	Actor-level attributes	Acknowledges agency; compatible with constructivist notion of context and place	Ignores structure; spatial models aggregate actor measures to spatial units as proxy for actor
'Friends and neighbors' effect (weak context, Cho 2003)	Micro-scale focus but scale depends on set of relations/ other actors	Actor-level attributes plus information from associated actors	Acknowledges social influences on individual behavior	Not interested in connections beyond focal actor
Context-as-place (Agnew 1987; Agnew 1996)	Localized	Place attributes	Acknowledges both within place and beyond place influences on behavior	Suited for qualitative interpretation/ few formal principles
'Regional context' (Flint 2002)	Variable but analyst defined (e.g. spatial weights matrix); fixed thereafter	Place attributes plus information from neighboring places (spatial lag)	Acknowledges regions of shared influences or interactions	Can be atheoretical; fixes interaction between units at a single scale
Embeddedness (Flint et al. 2009)	Variable but defined by extent of entire network	Interactions between actor pairs (relational dyads)	Accommodates multiple scales of actors and spatial connections (when formalized as a network)	Actor and place attributes are difficult; limited to dyad-based statistical models

scholars are well-positioned to make. Indeed, some of these techniques already used to illustrate context fall clearly within the GIScience tradition (e.g. site and situation per Goodchild 2018), while others lie just beyond or along emerging intellectual frontiers in geography and GIScience (e.g. networks and social network analysis). So long as the unit of analysis is simultaneously social and spatial, geographers will have much to add.

On this basis, this paper utilizes a network-based conceptualization of context that draws general inspiration from the previous efforts discussed above and is specifically derived from three interrelated propositions. First, following from King, Agnew, Goodchild, and others, context is always partly compositional and analytic methods must be able to accommodate attribute information about both social actors and the immediate settings they inhabit. Second, following from Agnew and others, context is always partly social and therefore relational. As such, contextualized methods must also accommodate interactions between actors across various scales. However, following from Flint, the underlying spatial distribution of these relations must also be taken into account. Third, following from Hess and Grossberg, context is also partly a function of larger systems of interaction and meaning that individual actors and places are always embedded within must also be taken into account. Together, these three propositions (summarized in Table 2) form the basis for a working theory about context that can be operationalized for quantitative research.

In the next section, I employ social network-based methods to illustrate this approach to context for a qualitative analysis. First, I introduce a network-based statistical model

Table 2. Following Hess (2004) and Grossberg (2013), I propose three elements that compose context for quantitative modeling. The relational element requires network-based modeling techniques which can easily accommodate spatial relations.

Element	Scale	Data type
Compositional context	Individual actor and/or place focus	Actor- and/or place-specific attributes
Relational context	Sets of actors and places	Social relationships between pairs of actors (dyads) and spatial relationships between places (geolocated actors)
Systemic context	Entire larger systems composed by actors and places	Whole system-level properties

that estimates the likelihood of a relationship forming between any two actors. Next, I draw on a dataset with information about a set of actors and their social and spatial relationships to develop a series of models that reflect the tripartite concerns for context described above. As I show, the statistical model that meets all of the above criteria is also the one that embraces all aspects of context while also capturing key properties of the observed network. While the methods involved were done outside of a GIS, it is a fundamentally geographic approach to the problem which reflects both the typical 'site and situation' approach to context in GIScience while also providing a basis to discuss context not just theoretically but analytically as well.

Exponential random graph models

In the last decade, the idea of combining geographical and network theory has gained significant attention. A new generation of works now integrates how the location of social actors in space, the spatial arrangements of social networks, and various spatial representations of the above influence social ties and, reciprocally, how social ties can contribute to form spatially embedded communities (Wong *et al.* 2006, Ter Wal and Boschma 2009, Adams *et al.* 2011, Bathelt and Glückler 2011, Daqing *et al.* 2011, Arentze *et al.* 2012, Gelernter and Carley 2015, Glückler and Doreian 2016). Further, social network-based data and methods are an emerging interest in geography (Radil *et al.* 2010; Sui and Goodchild 2011; Emch *et al.* 2012) and a GIS is often used to provide a spatial element to networked data (e.g. Gerber *et al.* 2013). These gradual rapprochements point to a deeper integration of social networks in geography and network-based techniques have potential to better understand some of the most fundamental concepts in geography.

Exponential Random Graph Models, or ERGMs, are a family of network-based statistical models that were developed to accommodate the inherent non-independence of network data. ERGMs also reflect an approach that goes beyond the descriptive methods that are conventionally used in social network analysis (Morris *et al.* 2008; see also Lusher *et al.* 2013). While ERGMs themselves are not yet common in geography, the model coefficients are interpreted in ways that are similar to more commonly used logistic regression models; ERGM coefficients are the log of an odds ratio that is meant to capture the probability of a relationship (hereafter referred to as a tie) forming between a pair of actors with a larger network, or a defined set of actors on a set (or sets) of relations that tie them together. More precisely, an ERGM is a tool for identifying how the characteristics of the members of a network, their various relationships with each other, and the 'larger social forces' that shape these relations 'can explain or predict the observed pattern of relationships' (Harris 2014, p. 5). ERGMs also represent an improvement on rather limited statistical network models that emphasize either actor characteristics or relations between actors but not both (van Duijn and Huisman 2011).

ERGMs are similar to some other tie-focused network models, such as multiple regression quadratic assignment procedure (MR-QAP), as both can accommodate whole network data. Another group of network statistical models are focused on a single actor's relationships with a group of other actors called alters; these models, are typically referred to as egocentric models, do not capture entire larger networks. However, ERGMs differ in that they can also accommodate actor attributes (which MR-QAP cannot) while they are also concerned with the characteristics of an entire network (unlike egocentric models). In this way, ERGMs are a relatively new statistical approach. Current implementations of the models are designed for binary networks (where a tie is either present or absent) although efforts are underway to extend the current estimation procedure to valued networks (Krivitsky 2012).

Estimating the likelihood of a tie between a pair of actors in a random network (meaning a network where the probability of tie between a given pair of actors is the same as for any other tie in the network) is relatively simple. The probability of a tie occurring under these conditions is the same as calculating the proportion of observed ties to all total possible ties; this ratio is the same as the descriptive measure of network density. However, actual social networks typically have properties that differ from the assumptions of random networks (Harris 2014). First, not all network actors have the same propensity to form ties and some actors will have many more ties than others (a network property called non-uniform degree distribution). Second, actors with similar characteristics or attributes form ties more often than those that do not (a network property called homophily). Third, among a trio of actors, ties tend to either form among all three or not at all (a network property called transitivity). In each case, the probability of tie formation is not independent of other ties, but conditionally dependent on ties with others (whether directly connected to a focal actor or not). These types of properties are not just functions of actor attributes nor of their relations. Instead, they are only observable by considering the entire system of actors and their relations as a whole. This is analogous to the third point about context in Table 2 and ERGMs have the capability of accounting for such complex dependencies in the estimation process.

Although the estimation process is mathematically complex and can be computationally extensive, the general form of the model itself is quite simple. An observed network that consists of a set of actors on a single type of tie between them is the variable to be explained (the network itself is analogous to the dependent variable in a traditional regression format) and a number of explanatory variables may be included to predict the probability of an additional tie forming in the original network (the dependent variable). In network terms, node-level and dyad-level variables can both be accommodated and evaluated against the probability of a tie formation. Further, network models can accommodate the location information of actors in physical space when this information is organized in network form (as is common in spatial regression models). In this way, location-based network data that is recognized by traditional spatial analytic approaches requires little to no translation (Radil et al. 2010).

Example – the case of world war I

Below I illustrate the use of an ERGM model with information about World War I, one of the most significant and most studied events in modern history (Sabol 2010). As an example, the war represents an ideal case to consider issues of context as it involved political conflicts at multiple spatial scales: below the scale of the state (for example, Serbian Black Hand nationalists), between limited groups of states within particular regions (such as competing territorial claims among Balkan states), and competition between 'great power' states/empires over far-flung colonies and global ambitions. Further, the war has been extensively studied qualitatively and quantitatively and its primary causes and drivers have been well documented. However, the rest of the discussion does not dwell on the details of the war nor seeks to develop the best possible model to evaluate a war-related research question. Instead, network-based data about the war is used to illustrate a broader point about ERGMs and context.

Helpfully for this goal, WWI was also the focus of several recent papers by scholars that used network analysis to model elements of the war. Flint et al. (2009) and Vasquez et al. (2011) examined how whole-network density measures change over time to explain the space-time spread of the war in the international system while Radil et al. (2013) approached the same task from the perspective of blockmodeling, or the grouping of actors into 'blocks' based on similar connection patterns. However, these efforts were descriptive rather than inferential and relied exclusively on relational data (data on the relationships between actors rather than about the actors themselves). Chi et al. (2013) took a different approach to blockmodeling, using network measures to create actor-level variables to use in a spatial regression model to explore the war joining process. All represented typical engagements with network methods – overly descriptive or laboring to translate relational data into a non-relational form. None addressed all the basic points about the context raised in the previous section.

Following a slightly altered network dataset from Radil et al. (2013), a network of 43 countries with 44 cases of war across the duration of the conflict (mid-1914 through late 1918) was used to fit a basic ERGM following Harris (2014) using the statnet package in R (Handcock et al. 2008). From this base model, I also estimated three other models that incorporate each of the three elements of context previously discussed. The base model began with a random graph that only predicted the probability of a new tie based on the observed number of ties with a network (the whole network property of density). In this case, a tie is the presence of war between two nodes; each node represents a country in the international system prior to the start of the war. A visualization of the war network itself is presented in Figure 1 which helps to clarify that while there were relatively few cases of war against all the possible cases of war (network density = 0.0487), war was clustered in the network, with just a few nodes possessing many ties to others.

The base model only incorporates a term that represents the ties in the war network. The equation below describes this model:

$$logit\left(P\left(Y_{ij} = 1 | n, \ Y_{ij}^c\right)\right) = \theta_{edges}\delta_{edges}$$

where θ_{edges} is the coefficient of the 'edges' term (equivalent to ties) and δ_{edges} is a change term that represents the addition of one additional tie between any pair of

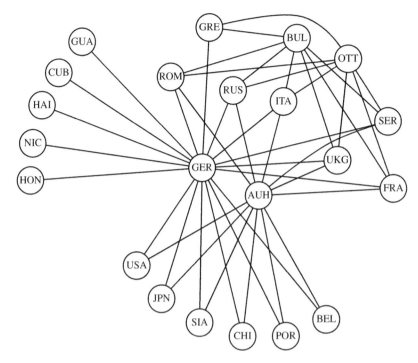

Figure 1. 23 out of 43 countries declare war between 1914 and 1918 and war is concentrated within the network, with a relatively few countries possessing many ties. The overall network density is 0.0478 which reflects that few of the possible total number of war ties are present. Countries without war declarations are omitted for clarity. Figure by author.

actors (n). Estimating this model in R returns the coefficient of the edges term and other information used in model assessment and related goodness of fit measures. Because the change term is based on the increase of a single tie in the unvalued or binary network, the change term is set at 1 when a tie is added (this means that the edges coefficient is always multiplied by one). The coefficient returns the log of the odds ratio that represents the probability of the addition of one tie (edge) to the network. In this instance, the log odds for the edges terms is −2.97158. Applying the following logistic function $\frac{1}{1+e^{-\left(\theta_{edges}\right)}}$ to this term calculates an actual probability of 4.87% of an additional war tie forming in this network (same as the overall network density). Using only the edges term, this model does not consider any compositional, relational, or systemic contextual influences but the main purpose of developing such a model is for use as a baseline for assessing fit as more complex models are built (Harris 2014).

The propensity for a few nodes to have a high number of connections and many nodes to have few connections is a common observation in many types of social networks (Harris 2014) and this is clearly the case in the war network as seen in Figure 1. Degree centrality is a node-level measure of the number of ties and the average nodal degree in this network is 2.05 which means that countries are at war with an average of two others during WWI. This propensity for uneven tie formation can be seen by comparing the degree distribution of the actual war network to the degree distribution of a simulated war network that has the same number of nodes and same density. The actual war network is to the left in

Figure 2 and a simulated network is to the right in the same figure. As the comparison suggests, tie formation is more complex than simply reflecting overall density and the ERGM model must account for such effects. A geographic interpretation of this propensity could be to consider how nodes are situated within place-based contexts and if place- or actor-specific attributes (referring to compositional context) influences degree distribution.

State attributes are included in the second model (compositional model) using military capability measures from the Correlates of War data (Singer 1987). These are measures of a state's share of the all military capabilities and therefore range from 0 to 1. To explore the issue, I recoded a state's capability score as either 0 for below the median for this set of actors and 1 for those above the median, then considered the patterns of tie distribution against these capability categories. Most war ties (26 ties out of 44 total, 60%) were formed between a higher-lower capability dyad. The fewest ties were formed between higher capability pairs (five ties, 12% of total ties) and the remainder were between lower capability pairs (13 ties, 28%). While the relationship between capability differential and war is a theme in the international relations literature, uneven power relations are also important in geography and can lead to a consideration of spaces in the international system where pairs of states with such capability imbalances are also within close proximity to each other.

As mentioned above, actor (or node-level) attributes can also be incorporated into the model to assess compositional effects on tie formation. These can be either categorical or continuous measures. To demonstrate, I included both a categorical and continuous variable: *cinc* is the military capability variable described above which ranges from 0 to 1 (included using the *nodecov* term) and *Europe* is a dummy variable that codes for whether a node is located in Europe, the epicenter of the war (included using

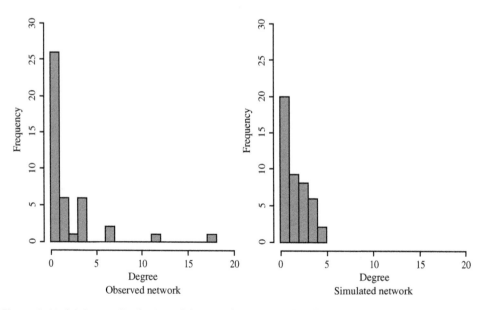

Figure 2. Nodal degree distribution of the actual war network (left) and a simulated network (right). Both networks have the same tie density, but the ties are distributed differently in the actual network, with some nodes having many more ties than others. Figure by author.

the *nodefactor* term). The calculation of probabilities is slightly different for these variables than for the *edges* term. For the continuous variable, the change statistic ($\delta_{nodecov}$) that is multiplied by the *nodecov* coefficient is the sum of the values for a dyad. For example, Germany's cinc measure is 0.1433 and Japan's is 0.0336. The summed score of 0.1769 would then be the multiplier for the *nodecov* coefficient reported below in Table 3. The multiplier is therefore different for each possible dyad pair and this must be kept in mind when using the model results to calculate the probability of tie formation. For the categorical *Europe* dummy variable, the change statistic multiplier ($\delta_{nodefactor}$) is slightly different still. If neither node in a dyad matches the dummy variable (or factor) used in the model (in this case it is 'Y' meaning the country is within Europe), the multiplier is 0. If one only one mode matches, it is set to 1, and if both nodes match the factor, it is set to 2.

Using the coefficients from the compositional model shown in Table 3 illustrates the point. The probability of a war forming between Germany (*cinc* = 0.1433, *Europe* = Y) and Russia (*cinc* = 0.1161, *Europe* = Y) would be calculated using the following formula:

$$P\left(Y_{ij} = 1 | n, \ Y_{ij}^c\right) = logistic\left(\theta_{edges}\delta_{edges} + \theta_{nodecov}\delta_{nodecov} + \theta_{nodefactor}\delta_{nodefactor}\right)$$

Substituting the coefficients from Table 3 with the correct change statistics gives:

$$P\left(Y_{Germany \ Russia} = 1\right) = logistic\left((-3.44 * 1) + (-3.16 * 0.1769) + (0.6 * 2)\right)$$

$$P\left(Y_{Germany \ Russia} = 1\right) = logistic(-2.8)$$

$$P\left(Y_{Germany \ Russia} = 1\right) = 0.057$$

This probability may seem low but it is higher than the base model probability of 4.87% and war is more likely to form between this pair than elsewhere in the network. Attributes are important to tie formation and contribute to the increased likelihood of the formation of war in the international system.

Because ERGMs estimate the probability of tie formation, it is necessarily a dyadic model meaning that actor attributes are always considered at the level of dyads rather than at nodes. As already discussed, continuous node-level attributes can be included

Table 3. Comparisons between the four models. AIC values indicated that the relational model was the best fit among the first three models. AIC is not reported for the systemic model as it is not recommended for use in comparing models with complex dependencies to those without them (Koskinen and Snijders 2013). * = p < 0.05.

	Estimate (SE)			
	Base model	Compositional model	Relational model	Systemic model
Edges (constant)	−2.97* (0.15)	−3.44* (0.30)	−3.84* (0.34)	−2.25* (0.32)
Actor attributes				
Capability (cinc)		−3.16 (2.90)	−4.24 (2.95)	−5.43 (19.24)
Europe (Y)		0.60* (0.22)	−0.53* (0.33)	0.47* (1.32)
Relations				
Contiguity			1.18* (0.38)	0.58* (2.42)
Rivalry			2.20* (0.48)	1.67* (2.76)
Systemic property				
GWD (degree)				−3.55 (1.09)
Fit				
AIC	353.7	350.3	313.2	

using the *nodecov* term but the change statistic multiplier is always calculated on the summed values for a dyad pair. This formulation means that it can difficult to assess the impact of a shift in value in a single node but this does allow for the consideration of homophily, or the propensity of node attribute similarity on a categorical variable to lead to tie formation. The change statistic only differs slightly from *nodefactor*; the multiplier is 1 if both node match the categorical variable and 0 if they do not. This is done in R using the *nodematch* term but is not demonstrated here.

ERGMs also can accommodate tie covariates in addition to actor attributes. As a practical matter, this means that one type of relationship (tie in a network) can be used to assess the probability of another forming. This is a critical capability of the model as allows the inclusion of relational covariates, not just actor attribute covariates. Helpfully, it also accommodates the type of spatial connectivity information used by spatial regression models (spatial weights matrix) but is not limited to this. To exemplify the second element of context (relationships between actors), I developed a relational model using two binary networks that have been identified as key to the initiation and spread of WWI: territorial contiguity (*ncontig*) and political rivalry (*nrival*). These are modeled as networks and included alongside the *cinc* and *Europe* dummy variables.

In the relational model, both the *nrival* and *ncontig* coefficients are positive and highly significant. All else held constant and in the presence of the attribute measures, the likelihood of war between two neighboring states is estimated to increase 3.3 times (odds ratio = 3.26, 95% CI = 1.52–6.95). Similarly, the model estimates that political rivals are 9 times more likely to go to war (odds ratio = 9.04, 95% CI = 3.47–23.63). The *Europe* dummy variable is no longer significant when these relational variables are included and was dropped from the final model given the presence of the *ncontig* network. The *cinc* scores were kept in for control purposes even though they too were no longer significant (see Table 3).

Model fit for ERGMs can be assessed though log-likelihood (LL), Akaike information criterion (AIC) or Bayesian information criterion (BIC) measures and used to compare models such as those demonstrated above (for details, see Harris 2014). AIC measures for each of the models are reported in ERGM summaries in R and included in Table 3. Another way to assess model fit is to examine how well a model captures important properties in the observed network which can be done through model simulation. Model simulation allows the comparison of a property with the observed network and can be based on any ERGM model, such as the ones developed so far. Using the last ERGM as an example, the observed WWI war network has 44 edges, but the simulated war network has 57 (see Table 4). Further, the number of states that do not fight (degree = 0) is 21; in the

Table 4. The simulated war network using the base model overestimates the number of wars, underestimates the number of countries that do not declare war, and distributes war ties throughout the network rather than reflecting the clustered form of the observed network. The systemic model demonstrates improvement on all issues.

Network	Ties	Actors with Degree = ___					
		0	1	2	3	4	5+
Observed WWI war network	44	21	5	6	1	6	8
Simulated war network (base model)	57	1	10	10	10	8	4
Simulated war network (systemic model)	48	8	9	10	9	3	4

simulated network, this only occurs once. This highlights that the model does a relatively poor job for accounting for the tendency for a few countries to have a large number of ties and for the large number of countries that don't fight at all during the war.

To account for the complexities of tie formation present in the observed network but missing from the ERGM-based simulation, other terms may be added to the model. These terms, which are referred to as 'dependence terms,' attempt to account for properties like uneven degree distribution described above. These terms require the use of a Markov chain Monte Carlo (MCMC) parameter estimation algorithm to calculate an approximate log-likelihood. The basic process of the MCMC estimation is to begin by selecting one network from a set of all possible networks, changing a tie, then comparing it to the original network to determine if it is a better fit (see Harris 2014 for more detail on this procedure). The details of this process are beyond the scope of this paper but it is important to note that the estimation process can exhibit problems with degeneracy, or the tendency of the MCMC simulated networks to be nearly devoid or nearly full of ties. Harris (2014) describes several MCMC parameters and settings that can be adjusted if degeneracy is observed.

Several terms have been proposed and developed that are meant to allow for the assumption of partial conditional dependence, or the assumption that dependence between ties that do not share a node in common may be due the presence of other ties in the network (see Hunter and Handcock 2006). In other words, tie formation is not just a property of attributes or even of dyads. The presence or absence of a tie can be partially due to ties in other, perhaps distant parts of the network. This is analogous to the idea of context being partly composed of large-scale structures who effects may be realized within a place but are largely the result of actors not present within a place. This is in keeping with the third point about context from Table 2 and is an important outcome toward the ability to model different elements of context. Importantly, the effects of these network properties can still be estimated as probabilities as the level of a dyad just as with the previous examples.

To demonstrate, I included one of the terms into the WWI ERGM that was suggested by Hunter and Handcock (2006). This term, called geometrically weighted degree (*gwdegree*), was designed to account for the decreasing degree distribution in observed networks, such as the WWI network. The term multiplies the frequency of each observed degree value by a weighting parameter and sums the values. The statistic is influenced by the proportion of high-degree node and a user-selected weighting parameter called a and can be specified in advance or estimated through the modeling process. The best specification/estimation of a is currently a source of debate in the ERGM literature but values of 0.25 and 0.7 have been suggested as providing reasonable results with networks with degree distributions similar to those in the observed WWI war network (see Koskinen and Snijders 2013, Harris 2014 for discussions of goodness of fit metrics for ERGMs). In the model below, the best value for a is estimated as a part of the MCMC simulation. Because this is simulation-based estimation, it would be important to specify the starting point to allow comparisons of competing models with such dependency terms. Problems with degeneracy may require adjusts to the default MCMC process, such as making the MCMC simulation longer than the default.

The final model summarized in Table 3 presents the results of an ERGM that uses the *gwdegree* term with $a = 0.77$ (this is reported in the summary by *gwdegree.decay*).

Although the term itself wasn't significant, the model provided a better estimate of degree distribution than did the previous models, a recommended way of evaluating the performance of ERGMs with complex dependencies. Simulating a war network using the probabilities estimated by the structural model provides also provides some insight. The simulated war network had 47 wars in the system (compared to 44 in the observed network) and degree counts that were far more like those in the observed network. When compared to the simulated base model, this structural model network also had more isolates (where degree = 0) and more high degree nodes. This is a step toward the properties in the observed network and a sign that some of the higher order dependencies present in such complex systems can be captured through these models. Although not demonstrated here, other dependency terms can be added to capture more properties that effect tie formation, such as triadic structures/transitivity. However, more dependency terms can dramatically increase the computational power and time required to complete the MCMC simulation.

Conclusion

As discussed in the beginning of the paper, context matters as a prime theoretical concern for geographers but engagement with the concept using quantitative modeling has been partial, even in the spatial analytic and GIScience traditions. To the extent that this issue has been addressed, it has largely relied on modeling spatial dependence at a particular scale (as in spatial regression models). As a consequence, context is formalized at one and only one scale that is preselected by the analyst. While this is an improvement over most efforts that either ignore context altogether or at least constrain it to micro-scale effects, context is formed through acts of agency at numerous spatial and social scales (Hess 2004; Grossberg 2013; Flint and Dezzani 2017). Building on the limited efforts to explicitly theorize context, I propose a three-tiered approach that includes actor- or place-specific attributes, relationships between pairs of actors or places (including spatial relationships), and larger-scale systemic effects that are not reduceable to either previous element.

When theorized in this way, ERGMs have potential as a meaningful inferential statistical tool to demonstrate the importance of all three elements of context. ERGMs can accept information/variables about actors themselves (attributes) and about the relationships between pairs of actors; the spatial relationships between these actors; and the larger systemic features that are not attributable to any specific actor or dyad, but that nonetheless affect the probability of relationships forming everywhere in a network (which also may be conceived of geographically as a region or area). Although ERGMs are not yet common in geography, they also represent a hopeful modeling pathway for those interested in more engagement with social theory and in engagements with a wider spectrum of the theories of how and why context matters. This is especially true given the increasing interest in social network-based concepts and techniques in GIScience.

Using the case of WWI, I developed a series of ERGMs to illustrate the potential to incorporate information both about individual actors and the relations that connect them, including spatial connections. Crucially, ERGMs allow for an evaluation of complex dependencies in a way that current spatial models do not. Rather than constraining unit interactions and, therefore the complex dependencies produced by these interactions, to a scale specified by the analyst, the likelihood of a tie forming between any two actors is created by the attributes of the actors themselves, their relations with others (social and spatial), and by the presence of other relational patterns that they themselves may not even directly be a party to. Most importantly, this type of model is a step closer to incorporating the complexities of context for quantitative analysis and represents a useful tool to help us advance the case for why context, and geography, indeed matters.

Disclosure statement

No potential conflict of interest was reported by the author.

ORCID

Steven M. Radil ⓘ http://orcid.org/0000-0002-0985-8930

References

Adams, J., Faust, K., and Lovasi, G.S., 2011. Capturing context: integrating spatial and social network analyses. *Social Networks*, 34 (1), 1–5. doi:10.1016/j.socnet.2011.10.007

Agnew, J.A., 1987. *Place and politics*. London: Allen and Unwin.

Agnew, J.A., 1996. Mapping politics: how context counts in electoral geography. *Political Geography*, 15 (2), 129–146. doi:10.1016/0962-6298(95)00076-3

Arentze, T.A., van den Berg, P., and Timmermans, H.J.P., 2012. Modeling social networks in geographic space: approach and empirical application. *Environment and Planning A*, 44, 1101–1120. doi:10.1068/a4438

Barnes, T.J., 2009. "Not only... but also": quantitative and critical geography. *The Professional Geographer*, 61 (3), 292–300. doi:10.1080/00330120902931937

Bathelt, H. and Glückler, J., 2011. *The relational economy: geographies of knowing and learning*. Oxford: Oxford University Press.

Chi, S.-H., et al., 2013. The spatial diffusion of war: the case of World War I. *Journal of the Korean Geographical Society*, 49 (1), 57–76.

Daqing, L., et al., 2011. Dimension of spatially embedded networks. *Nature Physics*, 7, 481–484. doi:10.1038/nphys1932

Deleuze, G. and Guattari, F., 1987. *A thousand plateaus: capitalism and schizophrenia*. trans. B. Massumi. Minneapolis MN: University of Minnesota Press.

Emch, M., et al., 2012. Integration of spatial and social network analysis in disease transmission studies. *Annals of the Association of American Geographers*, 102 (5), 1004–1015. doi:10.1080/00045608.2012.671129

Ethington, P.J. and McDaniel, J.A., 2007. Political places and institutional spaces: the intersection of political science and political geography. *Annual Review of Political Science*, 10 (1), 127–142. doi:10.1146/annurev.polisci.10.080505.100522

Flint, C., 2002. The theoretical and methodological utility of space and spatial statistics for historical studies: the Nazi Party in geographic context. *Historical Methods: A Journal of Quantitative and Interdisciplinary History*, 35 (1), 32–42. doi:10.1080/01615440209603142

Flint, C., *et al.*, 2009. Conceptualizing conflict space: toward a geography of relational power and embeddedness in the analysis of interstate conflict. *Annals of the Association of American Geographers*, 99 (5), 827–835. doi:10.1080/00045600903253312

Flint, C. and Dezzani, R., 2017. Defining and operationalizing context through a structural political geography for international relations. *In*: W.R. Thompson, ed., *Oxford research encyclopedia of politics*. Vol. 1. Oxford, UK: Oxford University Press, 429–450.

Fotheringham, A.S., Brunsdon, C., and Charlton, M., 2004. *Quantitative geography: perspectives on spatial data analysis*. Thousand Oaks, CA: Sage.

Gelernter, J. and Carley, K.M., 2015. Spatiotemporal network analysis and visualization. *International Journal of Applied Geospatial Research*, 6 (2), 78–98. doi:10.4018/ijagr.2015040105

Gerber, E.R., Henry, A.D., and Lubell, M., 2013. Political homophily and collaboration in regional planning networks. *American Journal of Political Science*, 57 (3), 598–610. doi:10.1111/ajps.2013.57.issue-3

Glückler, J. and Doreian, P., 2016. Social network analysis and economic geography—positional, evolutionary and multi-level approaches. *Journal of Economic Geography*, 16 (6), 1123–1134.

Goodchild, M.F., 2018. A giscience perspective on the uncertainty of context. *Annals Of The American Association Of Geographers*, 10–8 (6), 14761481.

Goodchild, M.F. and Haining, R.P., 2004. GIS and spatial data analysis: converging perspectives. *Papers in Regional Science*, 83 (1), 363–385.

Granovetter, M., 1985. Economic action and social structure: the problem of embeddedness. *American Journal of Sociology*, 91 (3), 481–510. doi:10.1086/228311

Grossberg, L., 2013. Theorizing Context. *In*: D. Featherstone and J. Painter, eds., *Spatial politics: essays for Doreen Massey*. Malden, MA: Wiley-Blackwell, 32–43.

Handcock, M.S., *et al.*, 2008. statnet: software tools for the representation, visualization, analysis and simulation of network data. *Journal of Statistical Software*, 24 (1), 1–29. doi:10.18637/jss.v024.i01

Harris, J.K., 2014. *An introduction to exponential random graph modeling*. Thousand oaks, CA: Sage.

Hess, M., 2004. 'Spatial' relationships? Towards a reconceptualization of embeddedness. *Progress in Human Geography*, 28 (2), 165–186. doi:10.1191/0309132504ph479oa

Hunter, D.R. and Handcock, M.S., 2006. Inference in curved exponential family models for networks. *Journal of Computational and Graphical Statistics*, 15 (3), 565–583. doi:10.1198/106186006X133069

King, G., 1996. Why context should not count. *Political Geography*, 15 (2), 159–164. doi:10.1016/0962-6298(95)00079-8

Koskinen, J. and Snijders, T., 2013. Simulation, estimation, and goodness of fit. *In*: D. Lusher, J. Koskinen, and G. Robins, eds., *Exponential random graph models for social networks: theory, methods, and applications*. New York, NY: Cambridge University Press, 141–166.

Krivitsky, P.N., 2012. Exponential-family random graph models for valued networks. *Electronic Journal of Statistics*, 6 (1), 1100–1128. doi:10.1214/12-EJS696

Kwan, M.-P., 2012. The uncertain geographic context problem. *Annals of the Association of American Geographers*, 102 (5), 958–968. doi:10.1080/00045608.2012.687349

Leenders, R. T. A. and Roger, T.A.J., 2002. Modeling social influence through network autocorrelation: constructing the weight matrix. *Social Networks*, 24 (1), 21–47. doi:10.1016/S0378-8733(01)00049-1

Lusher, D., Koskinen, J., and Robins, G., eds., 2013. *Exponential random graph models for social networks: theory, methods, and applications*. New York: Cambridge University Press.

Markusen, A., 1999. Fuzzy concepts, scanty evidence and policy distance: the case for rigour and policy relevance in critical regional studies. *Regional Studies*, 33 (9), 869–884. doi:10.1080/00343409950075506

Massey, D., 1993. Power geometry and a progressive sense of place. *In*: J. Bird, *et al.*, eds., *Mapping the futures: local cultures, global change*. London: Routledge, 75–85.

Massey, D., 1997. A feminist critique of political economy. *City*, 2 (7), 156–162. doi:10.1080/13604819708900068

Morris, M., Handcock, M.S., and Hunter, D.R., 2008. Specification of exponential-family random graph models: terms and computational aspects. *Journal of Statistical Software*, 24. http://www.jstatsoft.org/v24/i04/

O'Loughlin, J., 2000. Geography as space and geography as place: the divide between political science and political geography continues. *Geopolitics*, 5 (3), 126–137. doi:10.1080/14650040008407695

O'Loughlin, J., Flint, C., and Anselin, L., 1994. The geography of the Nazi vote: context, confession, and class in the reichstag election of 1930. *Annals of the Association of American Geographers*, 84 (3), 351–380. doi:10.1111/j.1467-8306.1994.tb01865.x

Pred, A., 1984. Place as historically contingent process: structuration and the time-geography of becoming places. *Annals of the Association of American Geographers*, 74 (2), 279–297. doi:10.1111/j.1467-8306.1984.tb01453.x

Radil, S.M. and Flint, C., 2013. Exiles and arms: the territorial practices of state making and war diffusion in post–Cold War Africa. *Territory, Politics, Governance*, 1 (2), 183–202. doi:10.1080/21622671.2013.814550

Radil, S.M. and Flint, C., 2015. A tale of two audacities: a response to Verweijen and van Meeteren. *Territory, Politics, Governance*, 3 (1), 112–117. doi:10.1080/21622671.2014.971629

Radil, S.M., Flint, C., and Chi, S.-H., 2013. A relational geography of war: actor–context interaction and the spread of World War I. *Annals of the Association of American Geographers*, 103 (6), 1468–1484. doi:10.1080/00045608.2013.832107

Radil, S.M., Flint, C., and Tita, G.E., 2010. Spatializing social networks: using social network analysis to investigate geographies of gang rivalry, territoriality, and violence in Los Angeles. *Annals of the Association of American Geographers*, 100 (2), 307–326. doi:10.1080/00045600903550428

Sabol, S., 2010. A brief note from the editor. *First World War Studies*, 1 (1), 1–2. doi:10.1080/19475021003621010

Sheppard, E., 2011. Geographical political economy. *Journal of Economic Geography*, 11 (2), 319–331. doi:10.1093/jeg/lbq049

Singer, J.D., 1987. Reconstructing the correlates of war dataset on material capabilities of states, 1816–1985. *International Interactions*, 14 (2), 115–132. doi:10.1080/03050628808434695

Sui, D. and Goodchild, M., 2011. The convergence of GIS and social media: challenges for GIScience. *International Journal of Geographical Information Science*, 25 (11), 1737–1748. doi:10.1080/13658816.2011.604636

Ter Wal, A.L.J. and Boschma, R.A., 2009. Applying social network analysis in economic geography: framing some key analytic issues. *The Annals of Regional Science*, 43 (3), 739–756. doi:10.1007/s00168-008-0258-3

Tita, G.E. and Radil, S.M., 2011. Spatializing the social networks of gangs to explore patterns of violence. *Journal of Quantitative Criminology*, 27 (4), 521–545. doi:10.1007/s10940-011-9136-8

van Duijn, M.A. and Huisman, M., 2011. Statistical models for ties and actors. *In*: J. Scott and P. J. Carrington, eds., *The SAGE handbook of social network analysis*. London: Sage, 459–483.

Vasquez, J.A., *et al.*, 2011. The conflict space of cataclysm: the international system and the spread of war 1914–1917. *Foreign Policy Analysis*, 7 (2), 143–168. doi:10.1111/fpa.2011.7.issue-2

Verweijen, J. and van Meeteren, M., 2014. Social network analysis and the de facto/de jure conundrum: security alliances and the territorialization of state authority in the post-Cold War Great Lakes region. *Territory, Politics, Governance*, 3 (1), 97–111. doi:10.1080/21622671.2014.912150

Warf, B. and Arias, S., eds., 2008. *The spatial turn: interdisciplinary perspectives*. Abington, NY: Routledge.

Wong, L.H., Pattison, P., and Robins, G., 2006. A spatial model for social networks. *Physica A*, 360, 99–120. doi:10.1016/j.physa.2005.04.029

Concluding comments

Yongwan Chun, Mei-Po Kwan, and Daniel A. Griffith

The seven papers in this book present challenges posed by uncertainty and geographic context in GIScience and geographic research that augment as well as go beyond the popular theme of error propagation (e.g., Arbia et al. 1998, 1999, 2002). Specifically, they cover the following three topics: (1) the impacts of uncertainty on spatial autocorrelation measures and the modifiable areal unit problem, (2) uncertainty in emerging new sources of geographic data and its impact on geographical analysis, and (3) the uncertain geographic context problem (UGCoP) that can lead to erroneous findings about individual behaviors and outcomes. These papers successfully address a wide range of uncertainty issues and propose novel approaches to addressing them. Nevertheless, researchers face more challenges and issues in uncertainty research that need further investigation in future studies (Griffith et al. 2015; Kwan and Schwanen 2018). The papers in this compendium suggest five subjects that require further investigation, contributing to a current well-established research agenda that already includes uncertainty explored through simulation experiments (e.g., Lee et al. 2018) and encountered by the use of imputations (Griffith and Liau 2020), massive satellite remotely sensed data (Griffith and Chun 2016a), and omitted variable surrogates (Griffith and Chun 2016b).

First, uncertainties in measurements have been largely overlooked in empirical geospatial analysis and spatial statistical modeling. One major reason for this oversight is the unavailability of uncertainty information for, especially, spatial survey data, although this lack of quantified precision has received some attention in the GIScience literature (e.g., Goodchild and Gopal 1989; Zhang and Goodchild 2002). The availability of margins of error (MOE) in the United States (US) American Community Survey (ACS) provides researchers with an opportunity to expand spatial analysis and modeling that incorporates uncertainty information, promoting a geography of error specialty area. However, how uncertainty in survey data can be appropriately incorporated into a descriptive equation specification such as regression models remains under-investigated. Classical independent observations sampling and measurement error conceptualizations furnish some guidance, as do existing uncorrelated data specification and stochastic error treatments. Nevertheless, explanatory variables that come from, for example, the ACS can have a non-trivial level of errors and, hence, such errors should be incorporated into a spatial statistical/econometric model specification.

Second, the impacts of uncertainty on spatial patterns merit further investigation. Research shows that ignorance of uncertainty could distort a spatial pattern manifestation of the corresponding variable, making a relevant spatial pattern difficult to discover (e.g., Sun et al. 2015). Jung et al. (2019a) extend this argument to spatial autocorrelation measures such as Moran's *I*. That is, Moran's *I* may fail to provide statistically reliable results when attribute uncertainty is not incorporated into its formulation. Recent research presents extended spatial autocorrelation measures in which uncertainty information is incorporated (Jung et al. 2019a; Jung et al. 2019b; Koo et al. 2019), including misspecification error (Griffith 2010), but this topic still needs further investigation for complicated situations such as multivariate spatial autocorrelation.

Third, the UGCoP furnishes a framework to investigate the "true casually relevant" geographic context that area-based observations rarely are capable of achieving (Kwan 2012a; Kwan 2018a). It allows an investigation of how differently individuals are affected by and/or respond to geographic contexts. Although the UGCoP has been adopted in geospatial health research, including environment exposure (Park and Kwan 2017; Wang et al. 2018), food access (Chen and Kwan 2015), obesity (Zhao et al. 2018), and mental health (Helbich 2018) studies, it has a great potential for addressing factors associated with other geographical phenomena, such as crime and selected residential relocation attributes (Kwan 2018a). A related and recently observed phenomenon called the neighborhood effect averaging problem (NEAP) also draws our attention to how individual exposures to contextual factors may vary in space and time in a highly complex manner (Kwan 2018b). From a technical point of view, advances in GIS, GPS, and location-aware mobile device technologies have made measuring different surroundings of individuals that lead to heterogeneous geographic contexts (Kwan 2012b) much easier. As a theme of future research, measurements of individual and real-time geographic context can be further extended to reflect more complex and heterogeneous conditions, including space-time variations (Park and Kwan 2017) and real-time exposure (Kou et al. 2020; Ma et al. 2020). In addition, analytical models should be developed to accommodate geographic contextual uncertainty in their specifications.

Fourth, as new emerging data sources increasingly have become available, assessment of their usability and reliability in GIScience and geographic research progressively has become important. Such data include those generated by georeferenced social media and volunteered geographic information (VGI) sources. Georeferenced social media data have been utilized in various research areas, such as urban structure (e.g., Chen et al. 2019), natural disaster management (e.g., Wang and Ye 2018), and food access (e.g., Widener and Li 2014). In parallel, potential problems about georeferenced social media data have been recognized, including the representativeness of their parent populations (e.g., Li et al. 2013). VGI provides a virtually free data source and has been popularly utilized. For example, Liu et al. in this book (Chapter 4) utilize OpenStreetMap data to investigate activity zones and mobility patterns. However, its quality is still a concern in terms of impacts on GIS applications and analysis results, even though the accuracy of VGI data has been investigated to some degree in the literature (e.g., Goodchild and Li 2012; Foody et al. 2013). For example, in this book, Delmelle et al. (Chapter 6) find discrepancies among network datasets from different VGI sources.

Finally, advances in uncertainty visualization would be beneficial to better explore uncertainty and subsequently to decide about a proper modeling specification. Uncertainty visualization has been investigated for decades in the literature (e.g., MacEachren 1992; Leitner and Buttenfield 2000; MacEachren et al. 2005; Koo et al. 2018). Such efforts have led to developments of visualization approaches, for example, the space-time aquarium (Kwan and Lee 2004) and volume rendering (e.g., Delmelle et al. 2014; Koo et al. 2020). In addition, researchers already have proposed approaches to incorporate uncertainty information in map classification for choropleth maps (Sun et al. 2015; Koo et al. 2017; Mu and Tong 2020).

However, uncertainty visualization still is limited because of its complex nature, with more information needing to be visualized than that associated with a conventional choropleth map. Also, a lack of uncertainty mapping tools in commercial GIS software packages constitutes another reason for its limited availability.

These five research topics are among those that provide rich investigation opportunities and allude to today's overarching big questions at the frontiers of geospatial uncertainty and context research. In addition, successful undertakings in these research areas can help establish a better understanding of uncertainty and context issues in GIScience and geography.

References

Arbia, G., Griffith, D. A., and Haining, R. (1998). Error propagation modelling in raster GIS: overlay operations. *International Journal of Geographical Information Systems*, 12, 145–167.

Arbia, G., Griffith, D. A., and Haining, R. (1999). Error propagation modelling in raster GIS: addition and ratioing operations. *Cartography & Geographic Information Systems*, 26, 297–315.

Arbia, G., Griffith, D. A., and Haining, R. (2002). Spatial error propagation when computing linear combinations of spectral bands: the case of vegetation indices. *Environmental and Ecological Statistics*, 10, 375–396; reply to commentary, 399–400.

Chen, X., and Kwan, M.-P. (2015). Contextual uncertainties, human mobility, and perceived food environment: the uncertain geographic context problem in food access research. *American Journal of Public Health*, 105(9), 1734–1737.

Chen, T., Hui, E. C., Wu, J., Lang, W., and Li, X. (2019). Identifying urban spatial structure and urban vibrancy in highly dense cities using georeferenced social media data. *Habitat International*, 89, 102005.

Delmelle, E., Dony, C., Casas, I., Jia, M., and Tang, W. (2014). Visualizing the impact of space-time uncertainties on dengue fever patterns. *International Journal of Geographical Information Science*, 28(5), 1107–1127.

Foody, G. M., See, L., Fritz, S., Van der Velde, M., Perger, C., Schill, C., and Boyd, D. S. (2013). Assessing the accuracy of volunteered geographic information arising from multiple contributors to an internet based collaborative project. *Transactions in GIS*, 17(6), 847–860.

Goodchild, M. F., and Li, L. (2012). Assuring the quality of volunteered geographic information. *Spatial Statistics*, 1, 110–120.

Goodchild, M. F., and Gopal, S. (Eds.). (1989). *The accuracy of spatial databases*. London: Taylor & Francis.

Griffith, D. A., (2010). The Moran Coefficient for non-normal data. *Journal of Statistical Planning and Inference*, 140, 2980–2990.

Griffith, D. A., and Chun, Y. (2016a). Spatial autocorrelation and uncertainty associated with remotely-sensed data. *Remote Sensing*, 8(7), 535.

Griffith, D. A., and Chun, Y. (2016b). Evaluating eigenvector spatial filter corrections for omitted georeferenced variables. *Econometrics*, 4, 29.

Griffith, D. A., and Liau, Y-T. (2020). Imputed spatial data: cautions arising from response and covariate imputation measurement error. *Spatial Statistics*, in press, https://doi.org/10.1016/j.spasta.2020.100419.

Griffith, D. A., Wong, D. W., and Chun, Y. (2015). Uncertainty-related research issues in spatial analysis in W. Shi, B. Wu, and A. Stein (Ed(s).), *Uncertainty modelling and quality control for spatial data*, 1–11. Boca Raton, FL: CRC Press.

Helbich, M. (2018). Toward dynamic urban environmental exposure assessments in mental health research. *Environmental Research*, 161, 129–135.

Jung, P. H., Thill, J. C., and Issel, M. (2019a). Spatial autocorrelation and data uncertainty in the American Community Survey: a critique. *International Journal of Geographical Information Science*, 33(6), 1155–1175.

Jung, P. H., Thill, J. C., and Issel, M. (2019b). Spatial autocorrelation statistics of areal prevalence rates under high uncertainty in denominator data. *Geographical Analysis*, 51(3), 354–380.

Koo, H., Chun, Y., and Griffith, D. A. (2018). Geovisualizing attribute uncertainty of interval and ratio variables: a framework and an implementation for vector data. *Journal of Visual Languages & Computing*, 44, 89–96.

Koo, H., Chun, Y., and Griffith, D. A. (2017). Optimal map classification incorporating uncertainty information. *Annals of the American Association of Geographers*, 107(3), 575–590.

Koo, H., Lee, M., Chun, Y., and Griffith, D. A. (2020). Space-time cluster detection with cross-space-time relative risk functions. *Cartography and Geographic Information Science*, 47(1), 67–78.

Koo, H., Wong, D. W., and Chun, Y. (2019). Measuring global spatial autocorrelation with data reliability information. *The Professional Geographer*, 71(3), 551–565.

Kou, L., Tao, Y., Kwan, M.-P., and Chai, Y. (2020). Understanding the relationships among individual-based momentary measured noise, perceived noise, and psychological stress: a Geographic Ecological Momentary Assessment (GEMA) approach. *Health & Place*, 64, 102285.

Kwan, M.-P. (2012a). The uncertain geographic context problem. *Annals of the Association of American Geographers*, 102(5), 958–968.

Kwan, M.-P. (2012b). How GIS can help address the uncertain geographic context problem in social science research. *Annals of GIS*, 18(4), 245–255.

Kwan, M.-P. (2018a). The limits of the neighborhood effect: contextual uncertainties in geographic, environmental health, and social science research. *Annals of the American Association of Geographers*, 108(6), 1482–1490.

Kwan, M.-P. (2018b). The neighborhood effect averaging problem (NEAP): an elusive confounder of the neighborhood effect. *International Journal of Environmental Research and Public Health*, 15, 1841.

Kwan, M.-P., and Lee, J. (2004). Geovisualization of human activity patterns using 3D GIS: a time-geographic approach. *Spatially Integrated Social Science*, 27, 721–744.

Kwan, M.-P., and Schwanen, T. (2018). Context and uncertainty in geography and GIScience: advances in theory, method, and practice, 108(6), 1435–1475.

Leitner, M., and Buttenfield, B. P. (2000). Guidelines for the display of attribute certainty. *Cartography and Geographic Information Science*, 27(1), 3–14.

Lee, M., Chun, Y., and D. Griffith. (2018). Error propagation in spatial modeling of public health data: a simulation approach using pediatric blood lead level data for Syracuse, New York. *Environmental Geochemistry and Health*, 40, 667–681.

Li, L., Goodchild, M. F., and Xu, B. (2013). Spatial, temporal, and socioeconomic patterns in the use of Twitter and Flickr. *Cartography and Geographic Information Science*, 40(2), 61–77.

Ma, J., Tao, Y., Kwan, M.-P., and Chai, Y. (2020). Assessing mobility-based real-time air pollution exposure in space and time using smart sensors and GPS trajectories in Beijing. *Annals of the American Association of Geographers*, 110(2), 434–448.

MacEachren, A. M. (1992). Visualizing uncertain information. *Cartographic Perspectives*, 13, 10–19.

MacEachren, A. M., Robinson, A., Hopper, S., Gardner, S., Murray, R., Gahegan, M., and Hetzler, E. (2005). Visualizing geospatial information uncertainty: what we know and what we need to know. *Cartography and Geographic Information Science*, 32(3), 139–160.

Mu, W., and Tong, D. (2020). Mapping uncertain geographical attributes: incorporating robustness into choropleth classification design. *International Journal of Geographical Information Science*, 1–21.

Park, Y. M., and Kwan, M.-P. (2017). Individual exposure estimates may be erroneous when spatio-temporal variability of air pollution and human mobility are ignored. *Health & Place*, 43, 85–94.

Sun, M., Wong, D. W., and Kronenfeld, B. J. (2015). A classification method for choropleth maps incorporating data reliability information. *The Professional Geographer*, 67(1), 72–83.

Wang, J., Kwan, M. P., and Chai, Y. (2018). An innovative context-based crystal-growth activity space method for environmental exposure assessment: a study using GIS and GPS trajectory data collected in Chicago. *International Journal of Environmental Research and Public Health*, 15(4), 703.

Wang, Z., and Ye, X. (2018). Social media analytics for natural disaster management. *International Journal of Geographical Information Science*, 32(1), 49–72.

Widener, M. J., and Li, W. (2014). Using geolocated Twitter data to monitor the prevalence of healthy and unhealthy food references across the US. *Applied Geography*, 54, 189–197.

Zhang, J., and Goodchild, M. F. (2002). *Uncertainty in geographical information*. London: Taylor & Francis.

Zhao, P., Kwan, M.-P., and Zhou, S. (2018). The uncertain geographic context problem in the analysis of the relationships between obesity and the built environment in Guangzhou. *International Journal of Environmental Research and Public Health*, 15(2), 308.

Index

Figures are indicated by italics, tables by bold type, and endnotes by an "n" and the footnote number after the page number e.g., 118n9 refers to footnote 9 on page 118.